Springer Theses

Recognizing Outstanding Ph.D. Research

For further volumes:
http://www.springer.com/series/8790

Aims and Scope

The series "Springer Theses" brings together a selection of the very best Ph.D. theses from around the world and across the physical sciences. Nominated and endorsed by two recognized specialists, each published volume has been selected for its scientific excellence and the high impact of its contents for the pertinent field of research. For greater accessibility to non-specialists, the published versions include an extended introduction, as well as a foreword by the student's supervisor explaining the special relevance of the work for the field. As a whole, the series will provide a valuable resource both for newcomers to the research fields described, and for other scientists seeking detailed background information on special questions. Finally, it provides an accredited documentation of the valuable contributions made by today's younger generation of scientists.

Theses are accepted into the series by invited nomination only and must fulfill all of the following criteria

- They must be written in good English.
- The topic should fall within the confines of Chemistry, Physics, Earth Sciences, Engineering and related interdisciplinary fields such as Materials, Nanoscience, Chemical Engineering, Complex Systems and Biophysics.
- The work reported in the thesis must represent a significant scientific advance.
- If the thesis includes previously published material, permission to reproduce this must be gained from the respective copyright holder.
- They must have been examined and passed during the 12 months prior to nomination.
- Each thesis should include a foreword by the supervisor outlining the significance of its content.
- The theses should have a clearly defined structure including an introduction accessible to scientists not expert in that particular field.

Liangliang Ji

Ion Acceleration and Extreme Light Field Generation Based on Ultra-short and Ultra-intense Lasers

Doctoral Thesis accepted by
Chinese Academy of Sciences, China

Author
Dr. Liangliang Ji
Institute for Theoretical Physics I
Heinrich-Heine-University Duesseldorf
Duesseldorf
Germany

and

Shanghai Institute of Optics
and Fine Mechanics
Chinese Academy of Sciences
Shanghai
China

Supervisor
Prof. Dr. Baifei Shen
Shanghai Institute of Optics
and Fine Mechanics
Chinese Academy of Sciences
Shanghai
China

ISSN 2190-5053 ISSN 2190-5061 (electronic)
ISBN 978-3-662-52434-3 ISBN 978-3-642-54007-3 (eBook)
DOI 10.1007/978-3-642-54007-3
Springer Heidelberg New York Dordrecht London

Printed on acid-free paper

Springer is part of Springer Science+Business Media (www.springer.com)

Parts of this thesis have been published in the following journal articles:

1. **L. L. Ji,** B. F. Shen, D. X. Li, D. Wang, Y. X. Leng, X. M. Zhang, M. Wen, W. P. Wang, J. C. Xu, and Y. H. Yu, "*Relativistic Single-Cycled Short-Wavelength Laser Pulse Compressed from a Chirped Pulse Induced by Laser-Foil Interaction*", **Phys. Rev. Lett. 105**, 025001 (2010).
2. **L. L. Ji,** B. F. Shen, X. M. Zhang, F. C. Wang, Z. Y. Jin, C. Q. Xia, M. Wen, W. P. Wang, J. C. Xu, and M. Y. Yu, "*Generating Quasi-Single-Cycle Relativistic Laser Pulses by Laser-Foil Interaction*", **Phys. Rev. Lett. 103**, 215005 (2009).
3. **Liangliang Ji,** Baifei Shen, Xiaomei Zhang, Fengchao Wang, Zhangying Jin, Meng Wen, Wenpeng Wang, and Jiancai Xu, "*Comment on 'Generating High-Current Monoenergetic Proton Beams by a Circularly Polarized Laser Pulse in the Phase-Stable Acceleration Regime'*", **Phys. Rev. Lett. 102**, 239501 (2009).
4. **Liangliang Ji,** Baifei Shen, Xiaomei Zhang, Fengchao Wang, Zhangying Jin, Xuemei Li, Meng Wen, and John R. Cary, "*Generating Monoenergetic Heavy-Ion Bunches with Laser-Induced Electrostatic Shocks*", **Phys. Rev. Lett. 101**, 164802 (2008).
5. **Liangliang Ji,** Baifei Shen, Xiaomei Zhang, Meng Wen, Changquan Xia, Wenpeng Wang, Jiancai Xu, Yahong Yu, Mingyang Yu, and Zhizhan Xu, "*Ultra-Intense Single Attosecond Pulse Generated from Circularly-Polarized Laser Interacting with Overdense Plasma*", **Phys. Plasmas 18**, 083104 (2011).

Supervisor's Foreword

The intensity of light has been greatly prompted to 10^{22} W/cm^2 due to the development of laser technology. Under such highly intense laser fields, light-matter interaction engages into the relativistic laser-plasma regime, presenting numerous nonlinear phenomena and significant potential for future applications. This thesis performs many-sided investigations on ultra-short (femtosecond, fs) and -intense (>10^{18} W/cm^2) lasers interacting with plasmas. Especially, we focus on laser-ion acceleration and new approaches of generating extreme light field.

Relativistic laser pulses can stimulate electrostatic fields in plasmas 3–4 magnitudes higher than those available in conventional accelerators. It is because of such feature that people hope to build high energy accelerators in a lab or even tabletop size, via laser-plasma interaction. Meanwhile, laser-plasma interaction also shows extraordinary potential in generating intense extreme light fields. Motivated by these prospects, researches on laser acceleration, high harmonic, and attosecond pulse (AP) generation have been developing for several decades. They are not only significant in critical applications such as energy source and medical treatment but also offer powerful tools to explore frontiers of fundamental physics. In Chap. 1, the basic properties of relativistic laser-plasma interaction, the main research topics and the current research states in the field are introduced.

In ion acceleration mechanisms, heavy ions are difficult to be accelerated because of their low charge-mass ratio. By particle-in-cell (PIC) simulations, it is found that when a relativistic circularly polarized (CP) laser interacting with a compound target, both light and heavy ions can be accelerated to the same velocity, which is higher than using pure heavy-ion target. Thus heavy ions being hard to be accelerated can be efficiently solved by mixing them to the light-ion target. Further, a "sandwich" micro-structured target is designed, which, after being accelerated by a 5×10^{19} W/cm^2, produces a carbon ion bunch with peak energy of 58 MeV and energy divergence better than 5 %. The scenario efficiently enhanced heavy-ion acceleration and improved the beam quality, thus is quite hopeful to be employed in future experiments. This efficient heavy-ion acceleration mechanism is described in Chap. 2.

Target thickness is a key issue in CP laser-driven light-pressure acceleration (LPA). Formal researches predicted a critical target thickness, below which all electrons will be pushed out of the target and disperse, presenting no stable acceleration. In Chap. 3, in-depth analysis shows that rising front of the laser pulse is crucial in LPA. For a gently rising pulse, the LPA scheme survives even the target is much thinner than the critical value. LPA is recently the most promising scheme of generating GeV ions. The above results clarified the key issue of critical thickness and relaxed the limit on target thickness. With a much thinner foil, ion peak energy can be increased by nearly one magnitude.

In Chap. 4, a new plasma approach of generating quasi-single-cycle relativistic laser pulse is proposed. In this proposal, the foil transparency to the incident CP pulse is nonlinearly modulated by laser intensity, resulting in a duration-compressed transmitted pulse. A quasi-single-cycle laser pulse with intensity above 10^{20} W/cm^2 is generated in the simulation. The new approach can produce ultra-intense and ultra-short pulse, which may prompt the research on intense single AP generation and laser wakefield acceleration. It also reveals that plasma is a powerful media of producing extreme light fields.

Generation of relativistic ultra-intense chirped pulse is introduced in Chap. 5. It employs dual CP lasers impinge interacting with a foil. The weaker pulse is reflected and highly chirped. Its spectrum width is enormously broadened and the central frequency is strongly blue-shifted. After dispersion compensation the chirped pulse can be compressed to short-wavelength single-cycled relativistic pulse. Further focused peak intensity reaches as high as 75,000 times as the initial pump laser! This plasma approach method generates broadband chirped pulse and therefore provides the possibility of obtaining light field close to the Schwinger limit and exploring the nonlinear physics of vacuum.

In Chap. 6, a scheme of generating single AP above 10^{21} W/cm^2 is proposed by CP laser being reflected from plasma target. A relativistic CP pulse can drive a plasma boundary to form a one-time violent oscillation. The incident pulse is reflected and severely compressed by the oscillating boundary, resulting in a single ultra-intense AP in the time domain. AP is widely concerned as an important tool to detect electron movements. Generating single-intense AP is all the more one of the top aims in high-field optics. The proposal also changed the traditional consensus in the "relativistic oscillating mirror" model that only linearly polarized laser could produce high-harmonics and attosecond pulses.

We hope the theoretical and simulative researches in this book may give some guidance on generating energy ions in experiments. Due to the great potential and prospect of generation of extreme light field by laser–plasma interaction, as shown in this book, we call for more attention. We thank Springer for their encouragement, and thank the staff of Springer for their support and patience.

Shanghai, 7 May 2013 Prof. Dr. Baifei Shen

Acknowledgments

First of all, I would like to thank my supervisor Prof. Dr. Baifei Shen, who has always been supportive and encouraging. His broad vision and intuition in physics have greatly inspired me. He noticed the prospect of laser acceleration very early and led me to this exciting frontier that is full of challenge and opportunities. He always works enthusiastically and keeps profound understandings about all aspects in laser-plasma physics. I could not have worked out each encountered difficulty without his detailed and patient guidance. Among all his enlightenments, I will always remember the one where he said, "Interests and projects are both important, but when they conflict, interests always go first."

I am grateful to Prof. Zhengming Sheng in Shanghai Jiaotong University and Prof. Wei Yu in SIOM, who prepared the letters of recommendation for me.

Many thanks to our group members Dr. Xiaomei Zhang, Dr. Fengchao Wang, Dr. Zhangying Jin, Dr. Xuemei Li, Dr. Meng Wen, Dr. Wenpeng Wang, Dr. Jiancai Xu and Ph.D. students Ms. Yahong Yu and Mr. Longqing Yi. I really enjoyed studying and discussing in this wonderful group. I would like to thank our secretary Ms. Shuyan Zhang, who helped me a lot in many paper works.

Most of all, I would like to thank my parents, who have been so supportive during my whole tour of study; my wife, Qiuping Zhang, who stood by my side and gave strong motivation for my Ph.D. study. This book is especially devoted to them.

Contents

Chapter 1
Introduction

Light-matter interaction is always one of the most critical topics in physics. Thanks to the invention of laser and development of laser technology [1–3], the light intensity has been enormously increased. Nowadays laser pulses with duration at the level of femtosecond (fs, 10^{-15} s) and peak intensity as high as 10^{22} W/cm^2 are available in laboratories. The laser field is so strong, way beyond the electric field inside atoms, that matter will be ionized immediately after being irradiated. The abundant ionized charged particles form an unbounded macroscopic state of matter [4], also known as the fourth state of matter–plasma. The interaction then enters the laser-plasma region. Plasmas show plenty of unique features under ultra-short and intense laser pulses. For example, in relativistic laser fields electrons oscillate transversely at the velocity close to the light speed thus relativistic effects must be counted in; electrons can be driven longitudinal by the laser ponderomotic force, to a velocity comparable to or even larger than the oscillating velocity. These characteristics induce colorful phenomenon in intense laser-plasma interaction, such as laser frequency shifting, high-order harmonic generation, plasma channel formation, relativistic solitons, electron/ion acceleration and so on. Most of the processes are highly nonlinear, which not only reveal the underlying principles in laser-plasma physics but also play important roles in variety applications. As it is, relativistic laser-plasma interaction is attracting more and more attentions and the related research has being one of the most promising parts of plasma physics.

1.1 Introduction to Relativistic Laser-Plasma Interaction

1.1.1 Development of Laser Technology and Corresponding Research Areas

In laser-matter interaction, when the laser is relatively weak, e.g., the laser field is below the Coulomb field of the outer-shell electrons in atoms, the bounded electrons would exhibit harmonic oscillations under the coaction of laser and Coulomb

L. Ji, *Ion Acceleration and Extreme Light Field Generation Based on Ultra-short and Ultra-intense Lasers*, Springer Theses,
DOI: 10.1007/978-3-642-54007-3_1,

fields, resulting in the Kerr nonlinearity. New nonlinear effects show up if both fields are comparable to each other, while the former is much higher than the latter, matter would be ionized in a short period, forming plasma. Figure 1.1 shows the history of laser development and the corresponding research regimes [5]. The invention of chirped pulse amplification (CPA) technique [1] promoted the laser output power from Giga-Watt (GW) level to Peta-Watt (PW), the pulse duration from picosecond (10^{-12} s, ps) level to fs, the focused intensity up to 10^{22} W/cm^2. Once such a highly intensive laser pulse impinges on the matter, electrons will be stripped from atoms in a laser period via field ionization. The process is now related to not only plasma physics but also the relativistic nonlinear optics.

In the near future, light intensity over 10^{24} W/cm^2 might be realized [5], where proton motion is also relativistic. Highly energetic nucleons colliding with each other may leads to nuclear fission or fusion, generation of mesons and neutrons, etc. The interaction then invades the laser-plasma nuclear physics regime.

When the light intensity is even higher, say higher than 10^{28} W/cm^2, electrons will be accelerated to obtain an kinetic energy equal to stationary energy m_ec^2 in a Compton wavelength ($\lambda_c = h/m_ec$). Here m_e is the electron mass at rest and c is the light speed in vacuum. In that case, electron-positron pairs are created, approaching the nonlinear quantum electrodynamics (QED) regime.

In this thesis, we focus on the interaction between plasma and fs laser with intensity over 10^{18} W/cm^2, i.e., the relativistic laser-plasma physics.

1.1.2 Typical Parameters and Characteristics

The immediate consequence of relativistic laser illuminating matter is that atoms will be ionized quickly. We start from estimating the inner electric field of an atom. Take hydrogen for example, the electric field is approximately

$$E_H = \frac{e}{r_0^2} = 5.1 \times 10^9\,\mathrm{V/cm}, \tag{1.1}$$

where $r_0 = 0.529 \times 10^{-8}\,\mathrm{cm}$ is the hydrogen Bohr radius. Recalling the relationship between laser intensity and amplitude of electric field

$$I = \frac{cE^2}{8\pi} = 1.33 \times 10^{-3}\,E^2(\mathrm{W/cm^2}), \tag{1.2}$$

where the field unit is V/cm, One would obtain the critical intensity corresponding to E_H

$$I_H = 3.5 \times 10^{16}\,\mathrm{W/cm^2}, B_H = 1.7 \times 10^7\,\mathrm{G}. \tag{1.3}$$

When the laser pulse is relativistic ($I_0 > 10^{18}$ W/cm^2), way beyond the ionization threshold, matter exists as plasma.

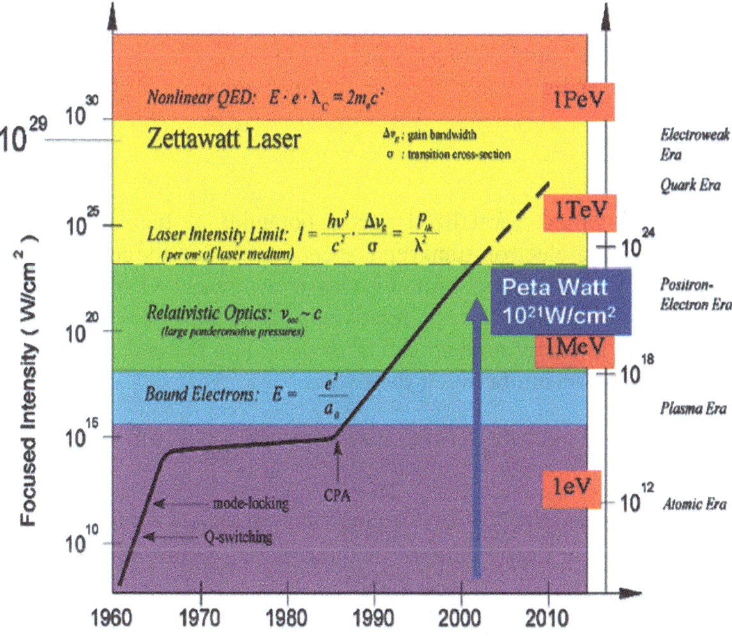

Fig. 1.1 Laser intensities versus time and the corresponding interacting regimes [5]

To further study laser-plasma interaction, some basic principles and parameters should be known, which will be introduced in the following sections.

1.1.2.1 Relativistic Threshold

The phrase "relativistic laser-plasma interaction" has been mentioned frequently. Here the definition of "relativistic" is given. It is related to the motion of free electrons in laser field. The momentum equation of an electron in an electromagnetic plane wave is

$$m_e \frac{d(\gamma \mathbf{v})}{dt} = e\left(\mathbf{E} + \frac{\mathbf{v}}{c} \times \mathbf{B}\right), \tag{1.4}$$

Here $\mathbf{E}$ and $\mathbf{B}$ are the electric and magnetic field of the wave; e is the fundamental charge; $\mathbf{v}$ is the electron velocity and $\gamma = 1/\sqrt{1 - v^2/c^2}$, respectively.

Considering only the transverse movement, i.e., momentum perpendicular to the laser propagating direction, the $\mathbf{v} \times \mathbf{B}$ force in Eq. (1.4) is neglected. The nonrelativistic ($\gamma = 1$) oscillating velocity is

$$v_\perp = -\frac{eA_0}{m_e c}\cos(k_0 x - \omega_0 t), \tag{1.5}$$

with $A_0 = cE/\omega_0$ the vector potential of light field, where ω_0 and k_0 are angular frequency and wave number of the laser field, respectively. The normalized amplitude of $v_\perp$ over light speed leads to a dimensionless variable

$$a_0 = \frac{eA_0}{m_e c^2}. \tag{1.6}$$

This is the so-called normalized vector potential of the laser field, which describes directly the electron transverse speed and hence the amplitude of laser pulse. As $a_0 << 1$, the electron motion is classic, while $a_0 = 1$ means the velocity amplitude is close to light speed and relativistic effect must be taken into account. For that the laser intensity corresponding to $a_0 = 1$ is defined as the relativistic threshold. The relationship between the two is

$$I_0\lambda^2 = \frac{\pi}{2} c A_0^2 = \left[1.37 \times 10^{18} \frac{\mathrm{W}}{\mathrm{cm}^2} \mu\mathrm{m}^2\right] a_0^2. \tag{1.7}$$

For laser wavelength of $\lambda = 1\,\mu\mathrm{m}$, the threshold intensity is about $1.37 \times 10^{18}\,\mathrm{W/cm^2}$. A useful hint is that the dimensionless amplitude a_0 is a Lorentz invariant.

Now let's look back to Eq. (1.5) and reserve the relativistic factor in the differential item at the left side. One obtains

$$\gamma = \sqrt{1 + a_0^2} = \sqrt{1 + \frac{I_0\lambda^2}{1.37 \times 10^{18}\,[\mathrm{W/cm^2}][\mu\mathrm{m}^2]}}. \tag{1.8}$$

On the other hand, since the oscillating velocity approaches the light speed, the $\mathbf{v} \times \mathbf{B}$ force becomes important and the motion parallel to laser propagating direction cannot be ignored any longer. A simple estimation gives $\mathbf{v} \times \mathbf{B} \propto E^2\hat{k}$, indicating that as the laser intensity increases ($a_0 > 1$), the longitudinal motion ($\propto a_0^2$) starts to surpass the transverse motion ($\propto a_0$) [6, 7]. As a result, when $a_0 \geq 1$ the electron trajectory exhibits a unique "8" shape, as seen in Fig. 1.2. The larger is a_0, the more obvious is the longitudinal motion.

1.1.2.2 Ponderomotive Force

In relativistic laser field, the longitudinal motion of an electron is important. However, if the light field is uniform both in space and time domains, as shown in Fig. 1.2, electron goes back to the initial position after one laser period, hence it presents no macro spacial displacement in a large time scale, acting like no longitudinal force existing. The situation is different in a non-uniform laser field. The integrations of Lorentz force over different laser periods are not zero any more, resulting in considerable macro motion. The electron seems to be pushed/pulled by

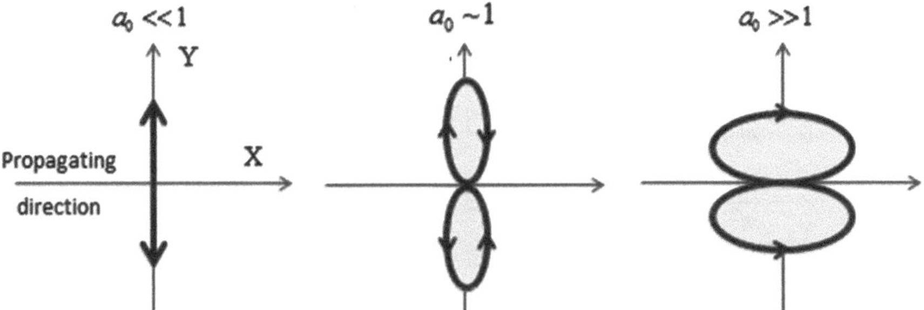

Fig. 1.2 Electron orbit under different laser amplitudes

a strong longitudinal force, named as the Ponderomotice force. We then deduce its accurate expression [4].

As the laser field oscillates with high frequency while the pulse envelop is slow-varying, it is convenient to split particle motion into two components: gradual movement of the oscillating center and fast oscillation around it, i.e., $\vec{r} = \vec{r}_0 + \vec{r}_1$. The non-relativistic equation of motion is

$$\frac{d^2\vec{r}}{dt^2} = \frac{q}{m}\left(E(\vec{r}) + \frac{1}{c}\frac{d\vec{r}}{dt} \times B(\vec{r})\right)e^{-\iota\omega t}, \tag{1.9}$$

Expanding the electromagnetic fields around the oscillating center, Eq. (1.9) can be expanded on different orders. The zero order generates oscillation solution

$$\vec{r}_0 = -\frac{q}{m\omega^2}\vec{E}e^{-i\omega t}, \tag{1.10}$$

which is exactly the results of Eq. (1.5). The first order of equation is

$$\ddot{\vec{r}}_1 = \frac{q}{m}\left[(\vec{r}_0 \cdot \nabla)\vec{E}\Big|_0 + \frac{1}{c}(\dot{\vec{r}}_1 + \dot{\vec{r}}_0) \times \vec{B}\right]. \tag{1.11}$$

Here the subscript ‘0’ denotes the corresponding value at the oscillating center. The time oscillating terms have been merged to $\vec{E}$ and $\vec{B}$. Taking average around the oscillating center gives $\left\langle \dot{\vec{r}}_1 \times \vec{B} \right\rangle = \dot{\vec{r}}_1 \times \left\langle \vec{B} \right\rangle = 0$, so that

$$\ddot{\vec{r}}_1 = \frac{q}{m}\left[\left\langle(\vec{r}_0 \cdot \nabla)\vec{E}\Big|_0\right\rangle + \left\langle\frac{1}{c}\dot{\vec{r}}_0 \times \vec{B}\right\rangle\right]. \tag{1.12}$$

Substitute Eq. (1.10) into the above expression, one have

$$
\begin{aligned}
\ddot{\vec{r}}_1 &= -\frac{q^2}{m^2\omega^2}\left[\left\langle(\vec{E}\cdot\nabla)\vec{E}\right\rangle + \left\langle\frac{1}{c}\dot{\vec{E}}\times\vec{B}\right\rangle\right] \\
&= -\frac{q^2}{m^2\omega^2}\left[\left\langle\nabla\frac{E^2}{2}\right\rangle - \left\langle\vec{E}\times(\nabla\times\vec{E})\right\rangle + \left\langle\frac{1}{c}\dot{\vec{E}}\times\vec{B}\right\rangle\right] \\
&= -\frac{q^2}{m^2\omega^2}\left[\left\langle\nabla\frac{E^2}{2}\right\rangle + \left\langle\frac{\partial}{\partial t}(\vec{E}\times\vec{B})\right\rangle\right] = -\frac{q^2}{m^2\omega^2}\nabla\frac{\langle E^2\rangle}{2} \\
&\equiv \frac{1}{m}\vec{f}_p
\end{aligned}
\tag{1.13}
$$

where the term

$$
\vec{f}_p = -\frac{q^2}{m\omega^2}\nabla\frac{\langle E^2\rangle}{2} \tag{1.14}
$$

is defined as the ponderomotive force acting on a single particle.

Several interesting features can be found about the ponderomotive force from the above expression: first, it is irrelevant to the charge polarity. Second, electrons experience much larger force than ions under the same laser condition. In normal occasions only electrons can be directly driven by laser field. Nevertheless, when laser is intense enough, say $a_0 \sim 1836$, ponderomitive force on protons should not be neglected. Third, the ponderomotive force tends to push charged particles from high field region to low field region as an electromagnetic pressure.

The fact that plasma being driven by laser through ponderomotive force induces macro effect of laser light pressure. One of the most favorite topics nowadays about laser-plasma acceleration is to employ laser pressure accelerating ions.

Another important aspect is that the ponderomotive force is strongly related to the laser polarization. The plane wave forms of different polarizations are
for linear polarization: $E = E_0(x)\sin(\omega_0 t)\hat{y}$;
for circular polarization: $E = E_0(x)[\sin(\omega_0 t)\hat{x} + \cos(\omega_0 t)\hat{y}]$

respectively. The corresponding ponderomotive forces according to Eq. (1.14) are then

for linear polarization:

$$
f_p = -\frac{e^2}{8m_e\omega_0^2}\nabla E_0^2(x)[1-\cos(2\omega_0 t)]; \tag{1.15}
$$

for circular polarization:

$$
f_p = -\frac{e^2}{4m_e\omega_0^2}\nabla E_0^2(x). \tag{1.16}
$$

One notice that the ponderomotive force of linearly-polarized (LP) laser contains two parts: a slow-varying part associated with the pulse envelope and a high frequency part that oscillates at twice the laser frequency, while the circularly-polarized (CP) only has the former. This critical difference introduces many interesting and important phenomena.

This thesis focuses on CP laser interacting with plasma, where one would see how its unique ponderomotive force plays the role.

1.1.2.3 Electron-Plasma and Ion-Acoustic Waves

Plasma, which is made of plenty charged particles, shows not only characteristics of free electrons/ions but, more importantly, collective effects. Various wave modes can be stimulated, where in relativistic laser-plasma interaction two main waves are mostly relevant: the electron-plasma and ion-acoustic waves.

Electron-plasma wave is also called plasma wave or Langmuir wave. It is the propagation of intrinsic collective oscillation originating from bulk charge separation, or plasma oscillation (Langmuir oscillation) for short. The intrinsic frequency is

$$\omega_{pe} \equiv \sqrt{\frac{4\pi n_0 e^2}{m_e}}. \tag{1.17}$$

It is a function of plasma density n_0, while in relativistic cases the electron mass should by multiplied by Lorentz factor γ of electrons. The electrostatic oscillation is localized and can propagate as a wave only if the plasma gets a temperature. In that case, the disperse relationship is found to be $\omega^2 = \omega_{pe}^2(1 + k^2\lambda_D^2)$, with λ_D the plasma Debye length.

Plasma wave is a kind of electrostatic wave, propagating with longitudinal charge separating field. In 1979, Tajima and Dawson proposed to use plasma wave accelerating electrons by injecting them into appropriate phases [8]. The maximum accelerating field available in a plasma wave can be calculated from the wave breaking condition of nonlinear Langmuir wave, which is [9–11]

$$E_p[\mathrm{V/cm}] = cm_e\omega_p/e \approx 0.96n_0^{1/2}\,[\mathrm{cm}^{-3}]. \tag{1.18}$$

Applying a typical plasma density of $n_0 \sim 10^{20}\,\mathrm{cm}^{-3}$ yields the electric field of $E_p \sim \mathrm{TV/m}$, which is four magnitudes higher than that of traditional accelerators (~ 100 MV/m). Since then, plasma-based acceleration brought new hope to realize table-top size high energy accelerators, and attracts more and more attentions.

The other important one is ion-acoustic wave. It is driven by thermal pressure and the ordinary dispersive equation is

$$\frac{\omega}{k} = \left(\frac{\gamma_i T_i + \gamma_e T_e}{m_i}\right)^{1/2} \equiv C_s. \tag{1.19}$$

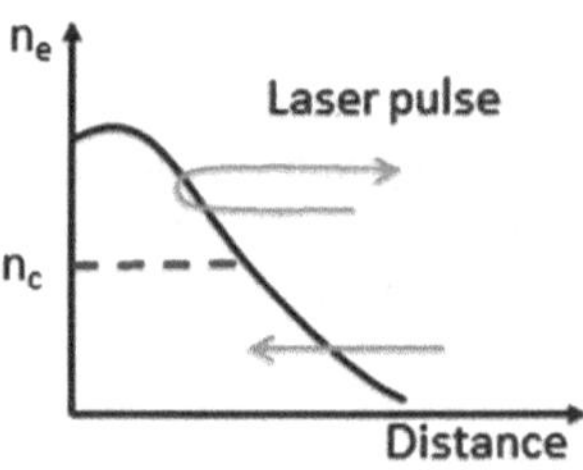

Fig. 1.3 Reflection of the incident laser pulse at the critical interface

Here γ_α and T_α are ratio of specific heat and temperature for certain species, respectively. The phase velocity is defined as the ion acoustic speed. Since electrons are normally with much higher temperature than ions, Eq. (1.19) is simplified to $C_s \sim \sqrt{T_e/m_i}$. Larger electron temperature means higher ion acoustic speed. The ion acoustic wave is closely correlated with plasma expansion and shock waves, which will be explained later.

1.1.2.4 Critical Plasma Density

According to the dispersion relation of electromagnetic (EM) wave in plasma

$$\omega^2 = \omega_{pe}^2 + k^2c^2, \tag{1.20}$$

The propagating phase velocity is larger with higher light frequency. When the light frequency equals that of plasma oscillation, the group velocity vanishes, suggesting that the light pulse cannot propagate in plasma any more. Instead, it will be reflected or absorbed. The plasma oscillating frequency represents the cut-off frequency for incident EM wave. Rewrite plasma density as a function of frequency from Eq. (1.17), one gets the critical plasma density corresponding to an incident laser with frequency of ω_L

$$n_c[\mathrm{cm}^{-3}] = m_e\omega_L^2/(4\pi e^2) = 1.1 \times 10^{21}/\lambda[\mathrm{\mu m}]_L^2. \tag{1.21}$$

When the plasma density is over the critical value or the plasma is overdense, i.e., $\omega^2 < \omega_{pe}^2$, the wave number turns to be imaginary. The incident EM wave decays in a certain length scale. Denoting $k = ik_s$, we have $k_s^2 = (\omega_{pe}^2 - \omega^2)/c^2$, so that the characteristic decaying length is $\lambda_s^2 = 1/k_s^2 = c^2/(\omega_{pe}^2 - \omega^2) \approx c^2/\omega_{pe}^2$, $(\omega << \omega_{pe})$. λ_s is the so-called skin depth of collisionless plasma. Incident laser is only allowed to penetrate into overdense plasmas by a length of λ_s and then reflected without dissipation, as shown in Fig. 1.3.

Self-focusing is another important effect in EM field propagating in plasma [6]. The refractive index of plasma, based on Eq. (1.20), is

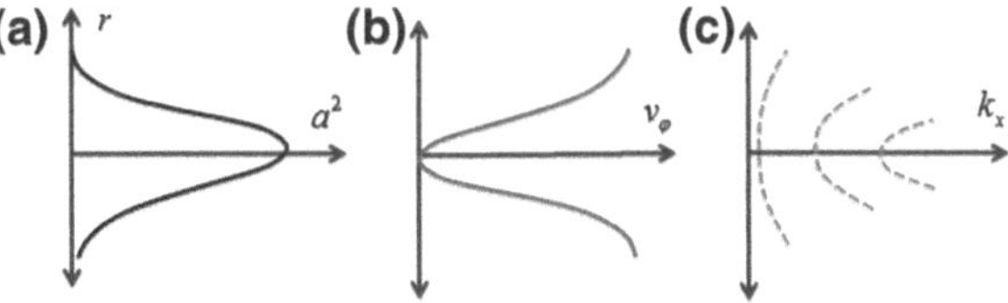

Fig. 1.4 Relativistic self-focusing mechanism

$$n = \frac{ck}{\omega_L} = \left(1 - \frac{\omega_{pe}^2}{\omega_L^2}\right)^{1/2} = \left(1 - \frac{1}{\gamma}\frac{n_0}{n_c}\right)^{1/2}, \tag{1.22}$$

where the relativistic effect has been considered. The refractive index is maximized at the axis due to two reasons: relativistic self-focusing, where laser intensity peaks at the axis and drops away from it, resulting in modulation of the electron relativistic factor; ponderomotive force, which expels electrons away from the axis and causes plasma channel. Both effects induce the refractive index decreasing from axis to abaxial positions. Plasma acts as a positive lens that focuses the propagating EM pulse, as shown in Fig. 1.4. Relativistic self-focusing appears when the laser power exceeds the critical power $P_c = 17(\omega_0/\omega_{pe})^2$ GW. On the other hand, light beams will be defocused by ionization. When self-focusing and defocusing compensate to each other, laser pulse could be self-guided in plasma and propagates for a long distance without deformation.

1.1.3 Theories and Simulation Methods

In this section, more rigid theoretical descriptions and the particle-in-cell simulation method will be introduced.

1.1.3.1 Maxwell Equations

To describe laser-plasma interaction, the scalar-vector-potential form of Maxwell equations is employed customarily [12]. Electric and magnetic fields **E** and **B** are defined by the vector and scalar potentials **A** and φ respectively

$$\mathbf{B} = \nabla \times \mathbf{A}, \tag{1.23}$$

$$\mathbf{E} = -\nabla\varphi - \frac{1}{c}\frac{\partial \mathbf{A}}{\partial t}. \tag{1.24}$$

Under Lorentz gauge

$$\nabla \cdot \mathbf{A} + \frac{1}{c}\frac{\partial \varphi}{\partial t} = 0 \tag{1.25}$$

Maxwell equations are rewritten to

$$\nabla^2\mathbf{A} - \frac{1}{c^2}\frac{\partial^2\mathbf{A}}{\partial t^2} = -\frac{4\pi}{c}\mathbf{J}, \tag{1.26}$$

$$\nabla^2\varphi - \frac{1}{c^2}\frac{\partial^2\varphi}{\partial t^2} = -4\pi\rho, \tag{1.27}$$

where $\mathbf{J}$ and ρ are densities of current and charge.

A more often-used gauge is Coulomb gauge or radiation/transverse gauge, which takes

$$\nabla \cdot \mathbf{A} = 0. \tag{1.28}$$

The equations are then

$$\nabla^2\mathbf{A} - \frac{1}{c^2}\frac{\partial^2\mathbf{A}}{\partial t^2} - \frac{1}{c^2}\frac{\partial}{\partial t}\nabla\varphi = -\frac{4\pi}{c}\mathbf{J}, \tag{1.29}$$

$$\nabla^2\varphi = -4\pi\rho. \tag{1.30}$$

After simplification and merging, Eq. (1.29) becomes

$$\nabla^2\mathbf{A} - \frac{1}{c^2}\frac{\partial^2\mathbf{A}}{\partial t^2} = -\frac{4\pi}{c}\mathbf{J}_\perp, \tag{1.31}$$

where $\mathbf{J}_\perp$ is the transverse electric current density. In many occasions, Eqs. (1.30) and (1.31) are more useful.

1.1.3.2 Lagrange-Hamilton Description

Here we introduce the Lagrangian and Hamiltonian of free charged particles in EM fields [13]

Recall particle kinetic equation including relativistic effect

$$\frac{d}{dt}\left(\frac{m\mathbf{v}}{\sqrt{1 - v^2/c^2}}\right) = q(\mathbf{E} + \frac{\mathbf{v}\times\mathbf{B}}{c}). \tag{1.32}$$

The Lagrangian is

$$L = -mc^2\sqrt{1 - v^2/c^2} - q(\phi - \frac{\mathbf{v}}{c}\cdot\mathbf{A}). \tag{1.33}$$

The canonical momentum is $\mathbf{P} = \gamma m\mathbf{v} + q\mathbf{A}/c = \mathbf{p} + q\mathbf{A}/c$, while the corresponding Hamiltonian turns out to be

$$H = \mathbf{P}\cdot\mathbf{v} - L = \sqrt{(c\mathbf{P} - q\mathbf{A})^2 + m^2c^4} + q\phi. \tag{1.34}$$

where $\mathbf{v} = \sqrt{\mathbf{v}_\perp^2 + \mathbf{v}_\parallel^2}$ and $\gamma = 1/\sqrt{1 - \mathbf{v}^2/c^2} = 1/\sqrt{1 - \mathbf{v}_\perp^2/c^2 - \mathbf{v}_\parallel^2/c^2}$.

For a plane wave with uniform transverse distribution, we have

$$\frac{\partial L}{\partial \mathbf{r}_\perp} = 0, \tag{1.35}$$

$$\frac{\partial L}{\partial \mathbf{v}_\perp} = \mathbf{p}_\perp + \frac{q}{c}\mathbf{A}_\perp = \text{const.}. \tag{1.36}$$

The Hamiltonian is exactly the total particle energy $H = E = \gamma mc^2$, so that

$$\frac{dE}{dt} = -\frac{\partial L}{\partial t} = c\frac{\partial L}{\partial x} = c\frac{d}{dt}\frac{\partial L}{\partial v_x} = c\frac{dp_x}{dt}, \tag{1.37}$$

$$\gamma mc^2 - cp_x = \text{const.}. \tag{1.38}$$

Free electron movement in uniform plane EM wave can be fully solved by Eqs. (1.36) and (1.38).

1.1.3.3 Equations of EM-Wave-Plasma Interaction

In the above single particle model, the charge and current effects on EM fields are both neglected, while dealing with plasma the huge electric charge and current should be coupled into the Maxwell equations. The Eqs. (1.30) and (1.31) are then upgraded to

$$\nabla^2 \boldsymbol{a} - \frac{\partial^2 \boldsymbol{a}}{\partial t^2} = n v_\perp, \tag{1.39}$$

$$\nabla^2 \phi = n - Z n_i. \tag{1.40}$$

Here Coulomb gauge is applied. All variables are normalized: $\mathbf{a} = e\mathbf{A}/m_e c^2$, $\phi = e\varphi/m_e c^2$, velocity by vacuum light speed c, time by ω_L^{-1}, length by k_L^{-1}, respectively, where ω_L and k_L are laser frequency and wave number. n and n_i are plasma electron and ion densities normalized by the critical density n_c. Z is the charge number of the ion species.

To supplement the equation set, the electron momentum equation is required

$$\frac{d\mathbf{p}}{dt} = \frac{\partial \mathbf{p}}{\partial t} + (\mathbf{v}\cdot\nabla)\mathbf{p} = -e\mathbf{E} - \frac{e}{c}\mathbf{v}\times\mathbf{B}. \tag{1.41}$$

Substitute Eqs. (1.23) and (1.24), one gets more compact relation after normalization

$$\frac{\partial}{\partial t}(\mathbf{p} - \mathbf{a}) = \nabla\phi - \nabla\gamma, \tag{1.42}$$

where $\mathbf{p} = \gamma \mathbf{v}$ the normalized momentum of electrons over c. We split the momentum into transverse and longitudinal components $\mathbf{p} = \mathbf{p}_\perp + \mathbf{p}_{||}$, where $\mathbf{p}_\perp = \gamma \mathbf{v}_\perp$ and $\mathbf{p}_{||} = \gamma \mathbf{v}_{||}$, satisfying $\nabla \cdot \mathbf{p}_\perp = 0$ and $\nabla \times \mathbf{p}_{||} = 0$, respectively. Equation (1.42) becomes

$$\mathbf{p}_\perp = \mathbf{a}, \frac{\partial \mathbf{p}_{||}}{\partial t} = \nabla(\phi - \gamma). \tag{1.43}$$

Here it is assumed that when $\boldsymbol{a} = 0$, the electron momentum is $\mathbf{p} = 0$, i.e., electrons are initially at rest.

Equations (1.39), (1.40) and (1.43) provide basic approach to describe laser-plasma interaction and explain quite a few phenomena very well. Many fruitful results have been obtained by solving them [14, 15]. Especially, for concerns of this thesis, the stationary solution is found about relativistic CP laser interacting with overdense plasma target [16].

1.1.3.4 Particle-In-Cell Simulation Method

Nowadays, computer simulations have become an independent subject other than theoretical discretion and experimental research, particularly in high field laser physics. On the one hand, related interactions are mainly nonlinear, increasing the complexity for analyzing, where one is not likely to see simple and clear physical images directly. On the other hand, experimental parameters are becoming more and more, which induces higher requirements on experimental setup. Revealing underlying principles by only comparing experimental and theoretical results would consume too much time, energy and costs. Using computer to simulate the real physical processes offers the perfect solution. It can not only show the clear physical images with much lower costs but also give experiments useful guides by pre-parametric studying, avoiding idle work. Also, it is a good support and supplement for experiments.

More importantly, people could add new mechanisms, use parameters not available in recent laboratories to perform researches in advance. Especially in the field of laser-plasma physics, it has played the role of "lantern tower" for several times.

The developed plasma simulating methods are classified as kinetic, fluid and hybrid simulations, respectively. Fluid simulation is to solve the simplified model or full set of magnetic hydrodynamic equations, which is more appropriate to study macro properties of plasma. While kinetic simulation focuses on particles in EM fields, more suitable for discovering the micro properties. It contains two approaches, solving the Vlasov or Fokker-Plank equation and the particle-in-cell (PIC) method.

Nowadays, PIC simulation is the mostly wide-used method in laser-plasma interactions, which is also the one employed in this thesis. In this section, we will give a brief introduction. In PIC method, plasma electrons and ions are pushed by the EM fields, which are distributed on each node of space cells.

$$\begin{aligned}\frac{d\mathbf{r}_j}{dt} &= \mathbf{v}_j,\\ \frac{d(m_j\mathbf{v}_j)}{dt} &= q_j\left(\mathbf{E}+\frac{\mathbf{v}_j\times\mathbf{B}}{c}\right),\\ &\oplus\text{Maxwell equations},\\ \rho_e &= \sum_j q_j\delta(\mathbf{r}_j-\mathbf{r}),\\ \mathbf{J} &= \sum_j q_j\mathbf{v}_j\delta(\mathbf{r}_j-\rho).\end{aligned} \tag{1.44}$$

Here q_j, m_j, $\mathbf{r}_j$, $\mathbf{v}_j$ are charge, mass, position and velocity of the particle with subscript "j", respectively. δ is the Delta function; $\mathbf{E}$, $\mathbf{B}$, ρ_e and $\mathbf{J}$ are electric field, magnetic field, charge density and current density, respectively.

In a PIC simulation, the simulating box is divided into many cells. Given positions and velocities of particles at a certain moment, the charge and current densities at every cell node are calculated to obtain the corresponding EM fields [17]. According to Eq. (1.44) the new positions and velocities can be deduced and so on. The calculating loop is as follows

In this method, the time step should be small enough to resolute the required problem. In most cases, the time step Δt is determined either by laser period or by plasma frequency

$$\Delta t \approx \frac{2\pi}{\omega_p} = \left(\frac{\pi m_e}{e^2 n_e}\right)^{1/2} \tag{1.45}$$

The spatial resolution Δx should also be sufficiently high to distinguish the scale of plasma collective behaviors. Normally this scale is comparable to the Debye length λ_{De} of plasma electrons. This means that the method is more suitable to simulate collisionless plasma, i.e., there is way more than one particle in the Debye sphere. For collisional plasmas PIC method is not a good choice, since it is difficult to compute with cell length smaller than the distance between each particle.

Another important requirement is the Courant condition. To keep the computation stable, the time step should be

$$\Delta t < \frac{\Delta x}{c} \tag{1.46}$$

The time step ($\sim 10^{-15}$ s) and cell size are usually quite small so that PIC simulations are applicable when the simulating time is much shorter than the life period of plasma and the plasma length scale should not be too large.

In reality, due to the limitation on computing ability the number of simulating particles is limited to $N_c \sim 10^6$. The real particle number is $N = nV$, where n is

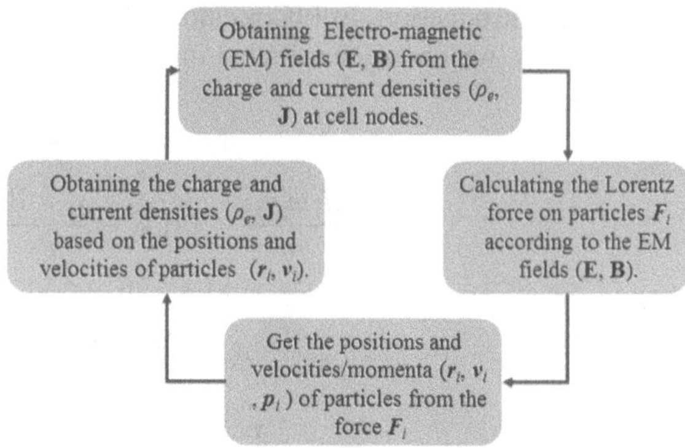

Fig. 1.5 The numerical loop for the particle-in-cell simulation method

particle density and V is plasma volume. For a typical lab laser-plasma system, $n \approx 10^{20}\,\text{cm}^{-3}$ and $V \approx 10^5\,\text{cm}^3$, i.e., $N \approx 10^{15}$, which is beyond the computing limit by many magnitudes. As a result, a simulating particle in the PIC method always represents many real particles. The above example gives mass and charge of a simulating particle 10^9 times of a real particle, hence in simulations

$$q_e = \left(\frac{N}{N_c}\right)e, m_e^{'} = \left(\frac{N}{N_c}\right)m_e, n_e^{'} = \left(\frac{N}{N_c}\right)^{-1} n_e. \quad (1.47)$$

Figure 1.5 describes the real three-dimensional process, which is sometimes complicated and would take too much computing sources. In lots of situations, a simplified low-dimensional system would be sufficient to show critical properties. Variables required with different dimensions can be found in [18].

1.2 Main Areas in Relativistic Laser-Plasma Interaction

This thesis focuses on relativistic laser-plasma interaction, i.e., laser intensity above 10^{18} W/cm^2. Hence researches in lower intensity regions will not be discussed. In the relativistic region, popular topics include laser "fast-ignition" fusion and particle acceleration. Strong high-order harmonics and generation of atto-second pulses, on the other hand, are attracting more and more attentions.

1.2.1 Laser "Fast Ignition" Fusion

Inertial confined fusion (ICF) normally uses high-power pulses, like laser pulses, electron beams or ion beams, to illuminate D-T pellets uniformly. The pellet surface is ionized and ablates in a very short period, forming high-temperature plasma which expands/explodes outward intensively. The induced counterforce produces extraordinary inward pressure, compressing the D-T plasma to an extremely dense and hot state, leading to nuclear fusions. Unlike in magnetic-confined fusion (MCF), plasmas in ICF do not require additional confinements. It relies on the inertial of the fuel and fusion is maintained before the plasma explodes away.

Traditional laser-ICF poses very strict conditions on spherical symmetry and uniformity of the irradiation, which is still a great challenge for nowadays technologies. Due to the invention of CPA technique, the laser intensity has been greatly prompted while the pulse duration is shortened. This leads to the proposal of "fast ignition" scheme by Tabak et al. [19, 20, 21]. The main idea is to separate the ignition and compression of the fuel, i.e., using an independent external trigger to ignite the pre-compressed fuel. An ultra-short laser pulse (duration $\sim 10^{-11}$ s) with intensity over 10^{20} W/cm^2 is focused on the surface of the deeply compressed fuel. The strong ponderomotive force would bore a hole in the critical interface and push it inward the highly dense core. Meanwhile, plenty of energetic MeV super-thermal electrons penetrate the critical interface, clash into the central region, and heat up ions immediately to achieve energy of 5–10 keV, realizing a fast ignition.

Fast ignition is quite hopeful to realize ICF. Although it still requires high compression rate to lower down the ignition energy, it does relax the demands on symmetry. There are three stages in a fast ignition: firstly, the fuel is compressed to reach several tens to hundreds of solid density; secondly, laser hole-boring produces a channel towards the highly dense region for the ignition laser; at last, a relativistic igniting laser pulse impinges on the plasma, generating energetic electrons, which deposit the energy to D-T ions to start nuclear fusion. Compared to traditional igniting method, fast ignition separates compression and ignition, leading to larger energy gaining and lower expense. In addition, the igniting source is alternative. Ion beams, for example, could also serve as drivers.

The critical issue in fast ignition is the energy transport of relativistic lasers through plasmas. The optimal case is that the incident laser intensity is well reserved after travelling through low-density plasma. Related effects in this process include relativistic self-focusing [22], plasma channel formation [23] and so on. Electrons can also be accelerated in plasma channel via direct acceleration [24] and random acceleration mechanism [25]. Another issue is the electron transportation in dense plasma. Recently, people used PIC and Hybrid simulations [26, 27] finding out the splitting and filamentaion of high-energy electron beams. Energy deposition in the ignition region is still an open question [28].

To avoid the nonlinear instabilities during laser propagating from the low-density region to high-density region, Kodama et al. [29, 30] employed a cone target, where the ignition laser goes through the hollow freely and hits the highly compressed center directly and ignites. The neutron yield is greatly enhanced, increasing the feasibility of fast ignition. Other methods such as using naturally formed cone channel [31] and energetic plasma beams [32, 33] have also proposed. Though with these supportive experiments, the physical process in fast ignition still requires more profound knowledge and further researches.

1.2.2 Electron Heating and Acceleration

1.2.2.1 Electron Heating

Electrons can get energy from laser fields easily via the following mechanisms

1. Resonance heating [34]: when a p-polarized laser pulse is obliquely incident on a non-uniform plasma, the ponderomotive force creates a low-density ramp near the critical surface and stimulates a plasma electrostatic wave, which traps and accelerates electrons by passing through the low-density region.
2. Vacuum heating [35]: oblique incidence of a p-polarized laser on a dense plasma generates a longitudinal electric field on the interface between plasma and vacuum, which drags electrons into vacuum and stimulates an electrostatic wave. Electrons are then heated up by the wave.
3. J × B heating [36, 37]: as been introduced in Sect. 1.1.2.4, laser field decays in overdense plasmas in a length of c/ω_{pe} (skin depth). For a linearly-polarized laser, electrons located in the skin layer experience the ponderomotive force of $f_x \propto (1 - \cos(2\omega_0 t))$, which leads to the heating of electrons in every $\lambda_0/2$ step along the incident direction. The heating process also reduces the plasma frequency and raises the skin length.
4. Random heating [38–40]: when electrons move in a plane EM wave, they won't obtain net energy after being time averaged. However if there is another EM wave propagating in different directions, electrons could be heated in a nearly random way once certain critical conditions are fulfilled.
5. Cyclotron resonance heating [24, 41]: in the low-density channel formed by laser hole-boring, the transverse and poloidal magnetic field could also heat electrons along the laser propagating direction.
6. "B-loop" heating [42]: in the laser hole-boring channel, if the width of the incident pulse is small, electrons would escape from the pulse focusing region from time to time. The toroidal magnetic field generated by the energetic electron flux then bends the escaping electrons back and keeps them in the interaction zone.

1.2.2.2 Electron Acceleration

Compared to laser heating electrons, people actually are more interested in accelerating electrons to gain collective velocity in a certain direction. The first mechanism is the direct laser acceleration (DLA) [43]. In a long plasma channel, electrons are expelled by light pressure, producing radial electric field, while the current inside generates poloidal magnetic field. The two fields together confine the relativistic electron bunch in the channel. Since they co-move with the laser pulse, electrons with appropriate phase will surf in the laser field, obtain oscillation energy and hence acceleration.

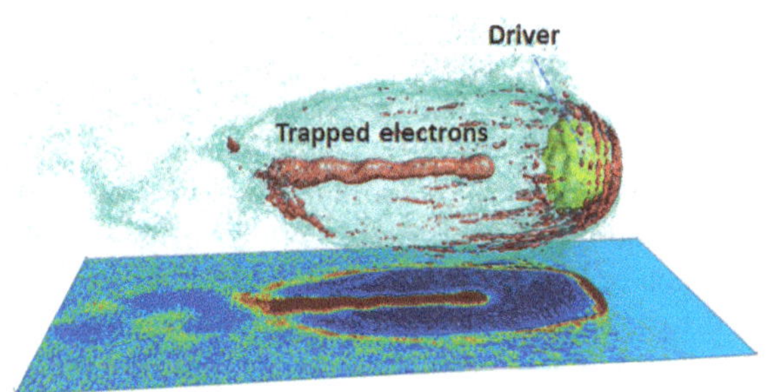

Fig. 1.6 Structure of a laser-driven bubble

The other mechanism, which attracts more and more attentions nowadays is the laser wakefield acceleration (LWFA) [43]. When an ultra-short and intense laser pulse enters an underdense plasma, the ponderomotive force expels electrons away from the axis. If the spatial length of the laser envelope is comparable to the plasma wavelength, a large-amplitude plasma wave could be stimulated behind the laser front, namely wakefield [10, 11, 44, 45]. Since the phase velocity of wakefield equals the laser group velocity, which is very close to light speed in most occasions, electrons trapped by the wakefield can be accelerated to extraordinary high energy. From Eq. (1.18), we know that accelerating gradient of the plasma wave is way beyond that of traditional accelerators, i.e., LWFA can obtain commeasurable energy in a greatly smaller distance. In such a manner, this new accelerating scheme brings the hope of table-tap size advanced accelerators.

Since Tajima and Dawson proposed the great LWFA idea in the year 1979, the scheme has made tremendous progress. In 2002, Malka et al. used 30 fs, 1 J laser pulse injecting to Helium gas with density of $(2\sim6)\times10^{19}$ cm^{-3}, an electron bunch with maximum energy of 200 MeV and 10^8 particles was observed. The energetic bunch is with small divergence but nearly 100 % energy spread.

Another breakthrough was in 2004, when groups from USA, France and U.K. realized self-injection of background electrons and obtained mono-energetic beams [46–49]. The three experiments all employed lasers of higher power (over 10TW) [50]. Such an intense pulse would become shorter during propagation, inducing large electron "bubble" [52, 53]. Plenty of self-injected electrons are trapped by the bubble field. After self-injection taking place, the wakefield transfers its energy to electrons thus prevents further trapping. A typical structure of a bubble is shown in Fig. 1.6, which displays an electron-free cave [55]. The ultra-short pulse blows all electrons out, which form the wall of the cave, leaving behind an ion background. The bubble is of the size similar to plasma wavelength.

Electron acceleration in the "bubble" or "blow-out" regime is somehow different from normal LWFA. It excites only one wakefield structure instead of a wave train. The accelerating field is also much more intense than the wave-breaking field [9, 10, 11], making it the most efficient acceleration mechanism at present. Electron beams at about 1 GeV with narrow energy spectrum and small divergence have been obtained [26, 48–54].

The main challenge lies in bubble acceleration is the stability. Interactions in this regime are highly nonlinear. Any inevitable changes of experimental parameters such as fluctuation of laser energy, variation of plasma density may lead to unstable acceleration. With certain laser-plasma parameters, the final obtainable energy is determined by the injecting point. The key of stabilizing acceleration is the accurate control of the injecting position. Using external injection or injection triggers turns out to be a good approach. In 2006, Faure et al. [56] took another relatively weak counter-propagating laser pulse colliding with the wakefield to control the injection position. This method had been theoretically proposed by Esarey et al. previously [57] and did greatly increase the stability. People could also use density gradient, ionization or pre-set nanowires [58] to manipulate injection.

Due to the development of LWFA, human beings are closer to the dream of producing ultra-energetic electrons in labs. The main tasks in the future are firstly raise the laser intensity to obtain higher electron energy and on the other hand, improve the beam quality, such as the energy spread, current, divergence and so on.

1.3 Laser Ion Acceleration

High energy ion bunches show extraordinary importance and vast prospects for numerous applications. They can be drivers of fast-ignition ICF, providing the new energy source; they can be used to treat cancers, offering the settlement of stubborn diseases for human beings; in high energy and particle physics, they collide with each other to produce fundamental particles, helping people to explore the underlying physic principles.

However the limit on accelerating gradient makes traditional accelerating techniques unbearable in the future, for their extremely high cost and occupying area. Since people noticed that ultra-intense lasers are able to stimulate electrostatic fields in plasmas several magnitudes higher than that in traditional ways, laser ion accelerators emerge as time requires. The goals of laser-ion-acceleration researches are firstly to increase ion energy as soon as possible and secondly to improve the energy spread. In some applications, the high total charge and good divergence are also greatly concerned. Several mechanisms have been proposed, from which target normal sheath acceleration (TNSA), electrostatic shock acceleration (ESA) and light sail acceleration (LSA) are approved by theoretical and experimental studies.

Since ion acceleration is one of the main topics in this thesis, it deserves a single section.

1.3.1 Target Normal Sheath Acceleration (TNSA)

TNSA is a relatively mature mechanism both theoretically and experimentally. As seen in Fig. 1.7, when a super strong laser ($\geq 10^{19}$ W/cm^2) interacts with a plasma target, electrons are heated up via mechanisms described in Sect. 1.2.2.1. They penetrate the target in a short period, forming a normal Debye sheath at back surface of the target. The sheath field is high enough to ionize surface atoms directly and pull the ions away from the target. In such a way, ions are continuously accelerated by the expanding electrostatic field. Eventually they emit along the normal direction of the back surface within a spatial angle [59–62].

In early experiments, laser pulses of intensity over 10^{19} W/cm^2 interacting with about 50 μm thick target generates 5–50 MeV ion beams [62, 63, 64], which move in the normal direction of the target back with good collimation (divergence angle <20°). Hatchett and Wilks et al. [59–62, 65, 66] confirmed from theory and simulations that these special beams do come from TNSA. Here we introduce the main analytical results.

The whole process of TNSA is actually similar to the expansion of thermal plasmas. We start from estimating the initial state of the hot plasma. Since interactions in TNSA usually take much longer time than the plasma oscillating period, it is reasonable to assume that the whole system is in thermal equilibration for electrons, which shows a Boltzmann distribution

$$N_{\mathrm{e,hot}} \sim \exp(-e\Phi/kT_{\mathrm{hot}}). \tag{1.48}$$

where T_{hot} it the electron temperature and Φ is the scalar potential. For relativistic laser intensities, the dominating heating mechanism is ponderomotive heating (or J × B heating), so that the temperature is approximately equivalent to the ponderomotive potential [67]

$$T_{\mathrm{hot}} \approx U_{\mathrm{ponderomotive}} \approx \left(\frac{I\lambda^2}{10^{19}\,\mathrm{Wcm^{-2}\mu m^2}}\right)^{\frac{1}{2}} \times 1\,\mathrm{MeV}. \tag{1.49}$$

Using the energy conservation law, one obtains the initial electrostatic field at the back surface of the target

$$E = kT_{\mathrm{hot}}/e\lambda_D. \tag{1.50}$$

here $\lambda_D = \sqrt{kT_{\mathrm{hot}}/4\pi e^2 N_{\mathrm{e,hot}}}$ is the Debye length. Let's take a typical set of parameters to see how intense this field could be. For $N_{\mathrm{e,hot}} \sim 2.5 \times 10^{19}\,\mathrm{cm^{-3}}$ and $T_{\mathrm{hot}} \sim 2\,\mathrm{MeV}$, the Debye length is $\lambda_D \sim 2\,\mu\mathrm{m}$, offering a sheath field of $E > 10^{12}$ V/m, which is far larger than the fields afforded by traditional accelerators.

Motivated by the initial sheath field, ions are accelerated and expand, exposing spatially exponential-decay distribution $N_{\mathrm{ion}} \sim \exp(-x/l_0)$ (x is the laser propagating direction). The plasma is quasi-neutral during expansion. Assuming

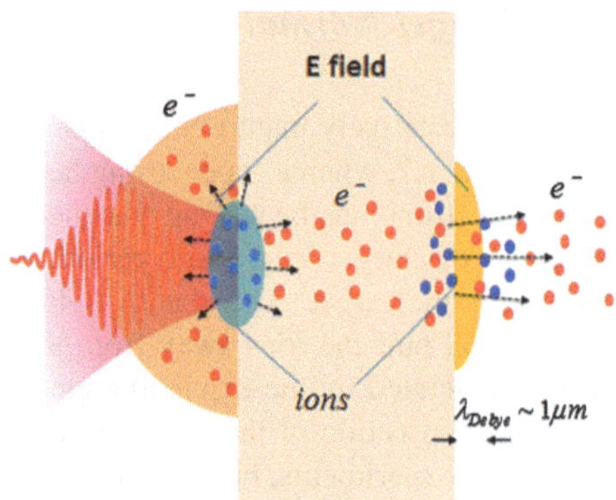

Fig. 1.7 Target Normal Sheath Acceleration

electrons are in isothermal equilibrium, the equations of continuity and motion about ions are

$$\frac{\partial n_i}{\partial t} + \frac{\partial (n_i v_i)}{\partial z} = 0, \tag{1.51}$$

$$\frac{\partial v_i}{\partial t} + v_i \frac{\partial v_i}{\partial x} = -C_s \frac{dn_i}{dx}. \tag{1.52}$$

where v_i is the ion velocity, $C_s = \sqrt{ZT_e/M_i}$ is the ion acoustic velocity, T_e is the electron temperature and M_i is the ion mass, respectively. Self-similar solution is found with quantities depending on $\xi = x/t$ instead of time and space independently

$$n_i = n_0 e^{-(1+x/C_s t)}, \tag{1.53}$$

$$v_i = C_s + x/t. \tag{1.54}$$

The accelerating field then is calculated through equation of electron motion

$$n_e e E = -\frac{\partial p_e}{\partial x}, \tag{1.55}$$

which gives

$$E = \frac{T_e}{eC_s t}. \tag{1.56}$$

With $l_0 = C_S t$ the length scale of the expanding plasma, we can find out that the field decays along with plasma expanding and electrons gradually cool down. Ion energy increases while the number of the most energetic ions drops. More accurate analysis can be found in [68].

The fact that TNSA originates from plasma thermal expansion induces a Maxwellian distribution of ion beams, i.e., with 100 % energy spread, which is not appropriate for many applications. To obtain mono-energetic ion bunches

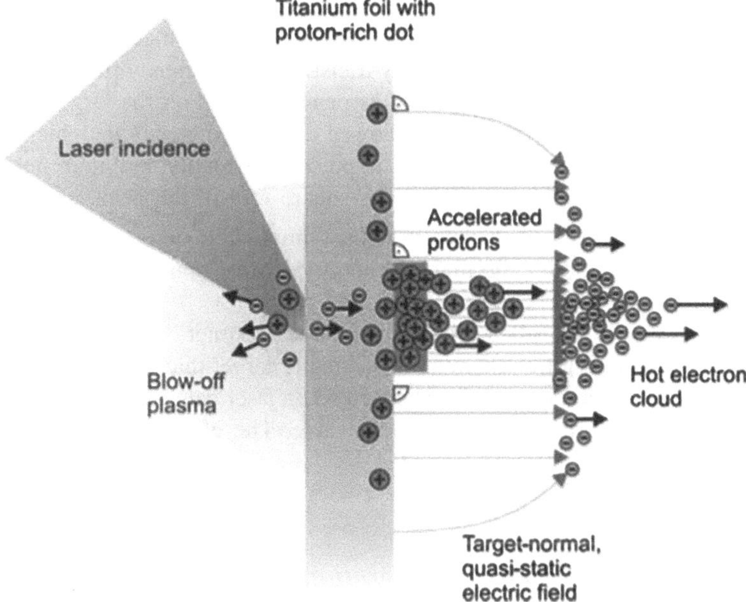

Fig. 1.8 Micro-structured double-layer target [72]

Esirkepov et al. suggested to use a double-layer target [69], with a small amount of light ions attached to a thick heavy-ion target. The latter provides huge number of thermal electrons affording the accelerating field while the light ions are preferentially accelerated, leaving heavy ions behind. As the attached light-ion layer is ultra-thin, all light ions experience almost the same sheath field and hence achieve similar energies in the end. As a result the energy spread is efficiently narrowed.

In 2006, Hegelich et al. applied the proposal in their experiments [70]. They attached a thin carbon layer (1 nm) to the back of a palladium target (20 μm) and then used a 20 TW/0.8 ps laser pulse hitting on the target. Mono-energetic C^{5+} beams with energy spread of 17 % were obtained. The average energy per nucleon is about 3 MeV and the longitudinal divergence is less than 10^{-6} eVs.

Bulanov et al. pointed out that transverse profile of laser intensity causes non-uniformity of TNSA laterally, which could also broaden ion energy spectrums [71]. To address this, Schwoerer et al. [72] employed a double-layer micro-structured target, as shown in Fig. 1.8. A micro hydrogen-rich target was placed behind a high-Z (Atomic number) foil. Since transverse size of the micro-target is much smaller than that of the sheath field, protons are accelerated near the axis, where the field is nearly homogeneous. MeV proton beam with energy spread of 20 % was produced.

Moreover, micro-lens-like technique was also adopted to select protons with similar energy from the Maxwellian distributed beam [73]. They got 6.25 MeV proton beam with spectrum width of 0.2 MeV by using a 5×10^{19} W/cm^2 laser.

The major advantage about TNSA is the relatively simple and convenient experimental setup. However limited by the accelerating mechanism it exhibits several shortcomings. Firstly, the energy transforming efficiency from laser to ions is low. People also tried using nanowires or multi-holes on the target surface to improve it [74, 75]. PIC simulations showed a triple efficiency increment, which has probably made the largest progress at present. Secondly and more fatally, the maximum energy by TNSA scales as the square root of the peak laser intensity $\sim I_0^{1/2}$. This weak scaling makes it almost impossible to generate GeV-level protons concerning one of the highest energy 58 MeV obtained by 3×10^{20} W/cm^2 laser [63].

An improved TNSA called "laser-breakout afterburner" (BOA) was proposed recently by Yin et al. [76], where a much smaller target thickness (10–500 nm) is necessary. The electron heating is enhanced when laser penetrates through the foil. In this way, GeV carbon beams are achievable. The efficiency is also raised to 5 %.

1.3.2 Electrostatic Shock Acceleration (ESA)

Shock is another highly nonlinear structure existing in fluid. Considering the propagating of a nonlinear acoustic wave, where the propagating velocity is higher for larger amplitude. The waveform is then distorted and finally resulting in a density discontinuity, the so-called shock wave [77]. In thermal-pressure driven shocks, the thickness of the shock front is at the level of molecular free path. Quantities such as velocity, temperature and density jump across the front. Shock wave plays an important role in ICF, via which the fuel can be compressed.

1.3.2.1 ESA in Plasma

In plasma, long-range EM force dominates, making dynamics more complicated. The ion acoustic wave is closely related to shock formation. It is a collective movement driven by electron thermal pressure. When the amplitude is sufficiently large, the nonlinearity steepens the waveform, aiming to cause discontinuity while plasma dispersion would disperse the waveform. When the two competing effects compensate to each other, the ion acoustic wave exists as a soliton. The ratio between the propagating velocity of the perturbation over ion acoustic velocity is defined as the Mach number $M = u_s/c_s$, where $c_s = (ZT_e/m_i)^{1/2}$ is the acoustic speed. Former theoretical studies [78, 79] indicate that when the Mach number is between 1 and 1.6, the solitary wave appears, as seen in Fig. 1.9b. There is a symmetric potential well in the soliton, where ions move back and forward and gain no net energy. Of course no reflection or acceleration shows up.

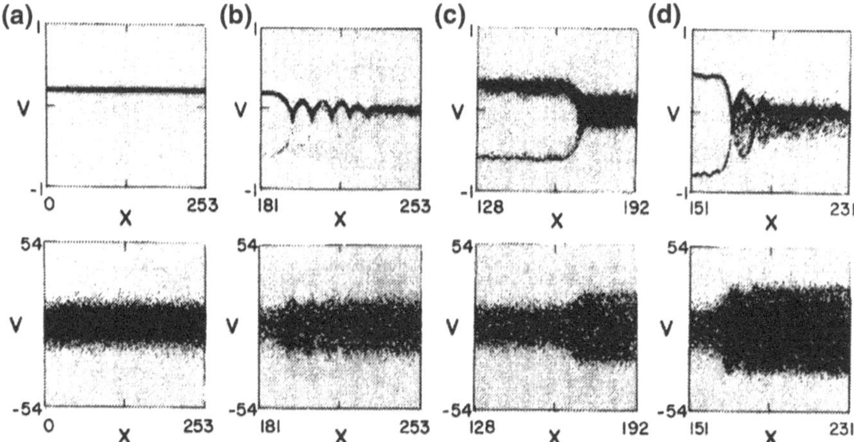

Fig. 1.9 The ion and electron phase space destitutions of $M = 1.3$, $T_e/T_i = 400$ at $t = 0$ (**a**) and $t = 298.2$ (**b**); of $M = 2$, $T_e/T_i = 535.3$ (**c**) and $M = 3$, $T_e/T_i = 400$ (**d**) [80]

However, as the Mach number is above 1.6, the nonlinear effect overcomes the dispersive effect. The discontinuity interface, or the shock wave, is created, as shown in Fig. 1.9c and d. In this case, a portion of local ions are reflected by the shock wave, destroying the symmetric solitary structure. The shock front moves at very high speed and is accompanied by strong electrostatic field, thus could be used to accelerate ions, namely electrostatic shock acceleration.

1.3.2.2 Light-Pressure Driven ESA

The laser intensity nowadays is so high that a relativistic laser could drive a shock wave directly via its light pressure. Though different from thermal-pressure driven mechanism, the Mach-number-condition (above 1.6) is also required. That is to say, the velocity of the moving interface between laser front and plasma should be slightly larger than ion sound speed.

When the laser is linearly polarized (LP), the stable component of the ponderomotive force would drive electrons forward. Meanwhile, the high-frequency oscillating part keeps heating electrons. The two mechanisms compete with each other. For one side, from Eq. (1.19) the higher is the electron temperature, the larger is the ion sound speed. For the other, stronger laser fields also provide higher driving velocities. It seems that laser-plasma parameters should be carefully selected to fulfill the Mach-number-condition. Experiential simulations suggest that higher laser intensity is more likely to stimulate a shock wave, which is reasonable because for more relativistic lasers, the electron longitudinal movement dominates the transverse one.

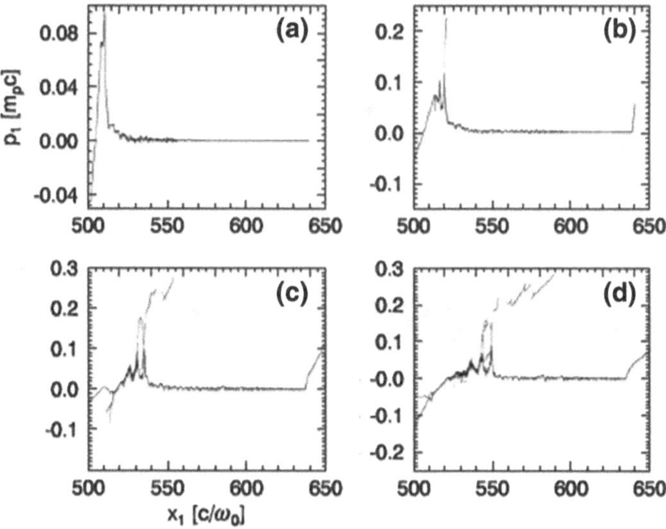

Fig. 1.10 Ion acceleration by LP-laser driven electrostatic shock [83]

Figure 1.10 reveals a typical phase-space evolution of ESA by a LP laser pulse [81–83]. One could clearly see that an electrostatic shock wave is launched from the target surface and propagates through the target. Some of the local protons are trapped, accelerated and reflected along the propagating direction. The reflected ions gain twice velocity as that of the shock wave. Considering momentum conservation the velocity is derived in the non-relativistic situation

$$\frac{v_i}{c} = 2\frac{u}{c} = 2\left(\frac{n_c}{2n_i}\frac{Zm_e}{M_i}\frac{I\lambda^2}{1.37 \times 10^{18}}\right)^{\frac{1}{2}}. \tag{1.58}$$

The ion energy is proportional to laser peak intensity. It should be noted that only a portion of ions are reflected and the acceleration is neither continuous nor stable. LP laser driven ESA [83–85] may be used to explain the generation of energetic cosmos rays but seems not appropriate for building accelerators.

After plenty of simulative researches, people realize that circularly-polarized (CP) lasers are more suitable for stable ESA. Since there is no oscillating component in the ponderomotive force, electrons are rarely heated, so that a highly nonlinear shock structure is easily stimulated. As shown in Fig. 1.11a, light pressure of the CP laser pushes electrons inward to form a dense skin layer, leaving ions behind. This would produce a strong charge-separating field. The laser pulse continually interacts with plasma and drives the shock structure forward. Local protons are successively trapped by the propagating shock field and reflected, leading to a high-energy ion beam. This is clearly displayed by the plateau phase space structure in Fig. 1.11b [86].

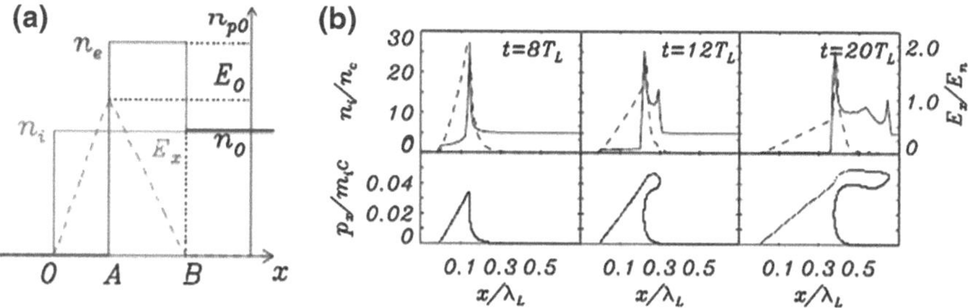

Fig. 1.11 Electrostatic shock acceleration driven by a circularly-polarized laser pulse [86]

In the following a simple dynamic model is introduced to deduce the velocity and amplitude of the accelerating field. Assuming a laser pulse with peak dimensionless amplitude a_L interacting with an overdense plasma target of density n_0, the light pressure pushes electrons inward while the charge separating field tries to pull them back. When the two forces reach a balance, the stable structure is built up. Without loss of generality we suppose a uniform density distribution of electrons in the skin layer. The electrostatic field is then

$$\begin{cases} E_x = E_0 x/d & \text{for } 0 < x < d\,, \\ E_x = E_0[1-(x-d)/l_s] & \text{for } d < x < d + l_s\,. \end{cases} \tag{1.59}$$

Where $d = |OA|$ is the snow-ploughed distance and $l_s = c/\omega_{pe}$ is the skin depth of the piled up electron layer. Together with the Poisson equation $E_0 = 4\pi e n_0 d$, charge conservation equation $n_0(d + l_s) = n_{p0} l_s$ and the balance relationship $E_0 e n_{p0} l_s \approx 2I/c$, one obtains the final ion velocity from the momentum equation

$$\frac{v_i}{c} = 2\frac{v_s}{c} = 2\sqrt{\frac{Z}{A}\frac{m_e}{m_i}\frac{n_c}{n_0}}\,a_L. \tag{1.60}$$

Equation (1.60) well explains the ion velocity scaling laws versus laser amplitude and plasma density in non-relativistic regime. The relativistic case is discussed in [87, 88]. Again the maximum ion energy is twice as the shock velocity. This scaling indicates that ion energy is proportional to laser intensity, which is more favorable than TNSA. One might notice that Eq. (1.60) suggests heavy ions, whose charge-mass ratio is small, are less efficiently accelerated. This is an intrinsic flaw for electric field accelerating. In Chap. 2, we will show a method to improve heavy-ion acceleration. Another deducing approach will also be given. Their conclusions are in coincidence.

Compared to LP lasers, CP lasers show many advantages in ESA. First of all, the acceleration is stable and continuous. Secondly, local ions are almost all reflected. The beam current and intensity are both very high. Lastly and most dramatically, the acceleration can be multi-staged. When all ions are reflected, the plasma bulk gains a collective velocity, which can still be driven by the laser pulse

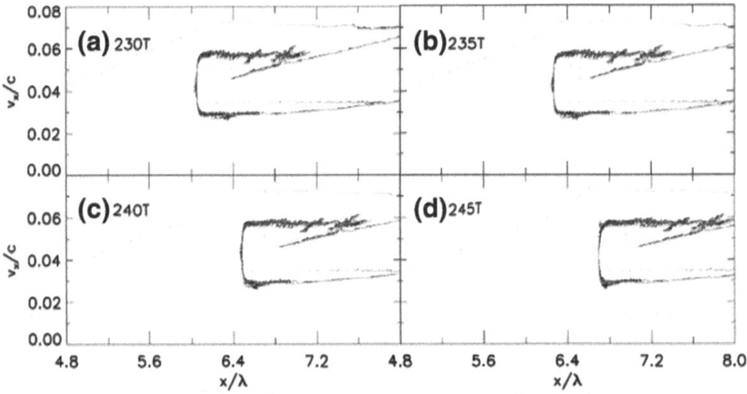

Fig. 1.12 Ion phase space (velocity versus position) distributions at different moments [89]

successively. This results in the second stage of ESA, and the ion velocity is doubled, as seen in Fig. 1.12. The interaction continues until the laser pulse deposits all the energy. In the end, ions experience a multi-staged acceleration [89]. Their final energy is several times higher.

Imagine that the target grows thinner; the time interval between two stages is smaller. When the target is thin enough so that the interval vanishes, ions are accelerated iteratively! It seems that the thin target, or the foil, is driven forward as a whole. We now reach one of the most efficient acceleration scheme-Light Sail Acceleration (LSA). Let's introduce it in following section.

1.3.3 Light Sail Acceleration (LSA)

People have noticed a long time ago that when illuminated by lights subjects feel light pressure and can even be pushed forward. A light beam reflecting from a moving boundary would generate a pressure of

$$p = \frac{2I}{c}\frac{c-v}{c+v}. \tag{1.61}$$

here I is the light intensity and v is the velocity of the interface. Equation (1.61) has accounted in the Doppler effect. An interesting fact is that light pressure is a Lorentz invariant, i.e., a constant in different inertial reference systems.

Someone used to suggest that light pressure could be employed to drive space ships. Nevertheless the light intensity and energy was quite low at that time, and the acceleration would be very rare. Due to the development of laser technologies, they both have been greatly improved that are able to generate extraordinary light pressure. Shen et al. [90] found in 2001 that a $a_0 = 100$ CP laser pulse could accelerate an overdense ($N_i = n_i/n_c = 10$) plasma foil as a whole. The efficiency

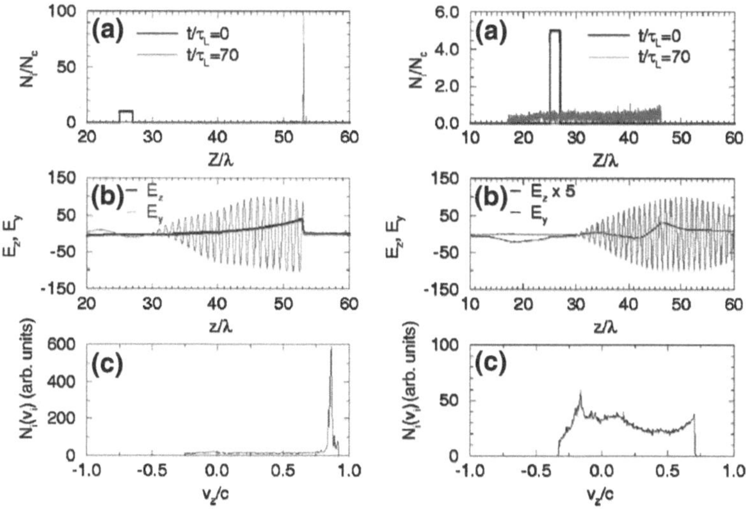

Fig. 1.13 Light-pressure acceleration of ions in the opaque (*left panel*) and transparent (*right panel*) regimes [90]

is quite high and the ion energy spectrum shows good mono-energetic feature, as displayed in Fig. 1.13.

The multi-staged acceleration evolves into LSA when the target thickness reaches the critical value. In 2008, Yan et al. [91] noticed that if the compressed electron layer is balanced by the light pressure and electrostatic force, LSA is obtained. They named it "phase stable acceleration (PSA)", for protons won't lose their phase during acceleration.

Analytically, the balance between light pressure and electrostatic force gives

$$a_L(1+\eta)^{1/2} \sim (n_0/n_c)(D/\lambda). \tag{1.62}$$

where D and η are target thickness and laser reflectivity; a_L and n_0 are normalized laser field amplitude and target density, respectively. To achieve PSA, one critical condition should be fulfilled: the ponderomotive force cannot overrun the electrostatic force of the charge separation field. Otherwise all electrons would be pushed out of the target and diverse. No balance or stable acceleration will be formed. We could simply assume that the ponderomotive act on electrons at the interface equals the Lorentz force. The condition is then

$$E_{//} = 4\pi e n_0 D > (v_e \times B_L/c) \sim E_L,$$

which can be written in a normalized form

$$a_L < (n_0/n_c)(2\pi D/\lambda). \tag{1.63}$$

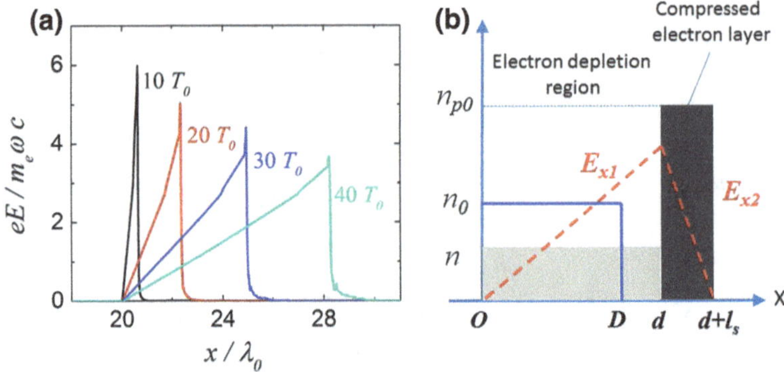

Fig. 1.14 The electric field at different stages in a PIC simulation (**a**) and the model of Phase-Stable Acceleration (**b**) [91]

In PSA, electrons are pushed inward by the light pressure and pile up to form a dense skin layer, as seen in Fig. 1.14b. The electron layer is driven forward while ions in the foil are accelerated by the positive charge separation field. Inside the skin layer, the field peaks at the left boundary and decays to zero at right boundary. This means ions located at the left side experience most intense acceleration so that they could catch up with ions in the front. The repeating process generates a loop structure in ion phase space. Ions always stay in the electron layer without losing their phase, that's why it is called phase-stable acceleration.

Macroscopically, the whole foil is driven by laser light pressure. Hence, the foil momentum should be [92]

$$\frac{d}{dt}(\gamma\beta) = \frac{m_i c}{2\pi n_0 D}\frac{1-\beta}{1+\beta}E_L^2(t-x/c). \tag{1.64}$$

here β is the foil velocity normalized on light speed c and $\gamma = (1-\beta^2)^{-1/2}$ is the corresponding relativistic factor. The electron momentum is neglected due to its much smaller mass. The right item in Eq. (1.64) is the light pressure including the Doppler effect. When solving the equation, one should notice the real laser field acting on the target is modified by foil motion.

LSA (or PSA) can generate ion beams with very low energy spread. In the one-dimensional simulation, a relativistic CP laser pulse of peak intensity $a_0 = 5$ and pulse duration 330 fs produced a proton beam with 300 MeV energy and only 4 % energy spread. The converting efficiency is over 20 %.

Actually, ultra-relativistic LP laser pulses can also accelerate ions in the LSA regime. As long as the LP laser field is strong enough, light pressure will dominate over thermal effect. In 2004, Esirkepov et al. [92] used $a_0 = 316$ LP laser pulse with sharp pulse front to accelerate an overdense foil, as seen in Fig. 1.15. They named it the laser-piston acceleration, where ions were accelerated to several GeV in a short period. Simulation results gave an energy efficiency above 40 %.

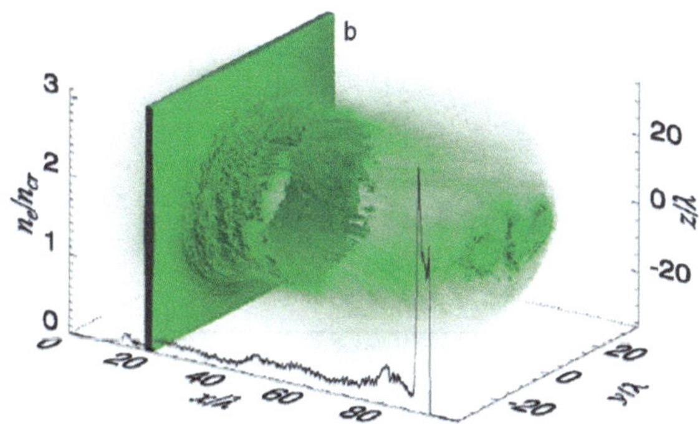

Fig. 1.15 Density distribution of foil ions in the laser-piston acceleration scheme [92]

LSA is very promising to produce GeV-level ion beams with higher efficiency and smaller energy spread than in TNSA. The mechanism has been proved by experiments [93]. However, there are still some difficulties to bring it further. Firstly, very high laser contrast is required to guarantee that the micro/nano-meter thin foil won't be destroyed by the pre-pulse. Secondly, it was found out recently that [36, 94, 95] various instabilities such as Rayleigh-Taylor-like instability and Weibel-like instability develop during interaction. The foil may also be destroyed before ions getting accelerated. How to restrain these instabilities become one of the central problems in LSA, and people are still working on it [96, 97].

1.3.4 Wakefield Ion Acceleration

Now we have LSA to generate GeV ions. What's the potential of plasma-based accelerators? Is it possible to produce several tens of GeV or even TeV ions with LSA? The answer is not so affirmative. As pointed out in [92], when the foil velocity becomes relativistic, ion energy increases slowly with time, just like $\sim t^{1/3}$. Further acceleration is not realistic because of the longer accelerating distance and time. On the other hand, the accelerating field is totally determined by the laser light pressure. One should note from Eq. (1.64) that once ions are relativistic, say protons with energy of 14 GeV, the light pressure would drop by three magnitudes. The accelerating field is also greatly decreased, making the mechanism not so efficient.

In 2007, Shen et al. [98] noticed that laser-driven wakefield could not only trap electrons but also ions in certain conditions. Trapped ions are efficiently accelerated by the positive wakefield, gaining several tens of GeV energy. Since wakefield is a very stable structure during propagation, the acceleration gradient will be maintained in a long distance, thus very promising to produce ultra-energetic ions.

To increase the trapping rate, some methods were also proposed, like combining LSA and LWFA [99–101]. In those proposals, an ultra-thin foil is presented before or in the low-density plasma region. Protons in the foil are firstly pre-

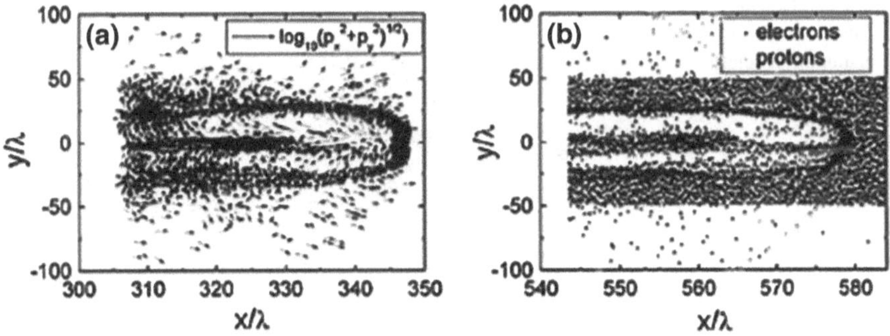

Fig. 1.16 Combined Light-pressure and wakefield acceleration. Distributions of the momenta of background electrons (**a**) and densities of electrons and protons (**b**)

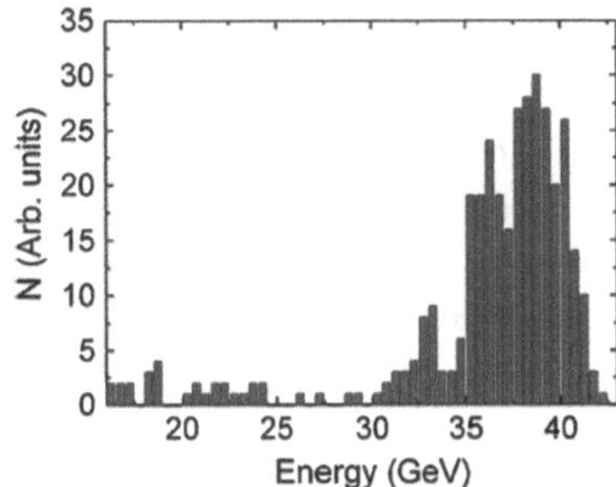

Fig. 1.17 Proton energy spectrum from wakefield acceleration [99]

accelerated by the laser light pressure and hence more easily trapped by the wakefield. As seen in Fig. 1.16b, foil protons are well confined in the bubble and experience long-term acceleration. Two-dimensional (2D) PIC simulation suggests that mono-energetic proton beam with peak energy of 38 GeV can be obtained using 10^{23} W/cm^2 CP laser pulse, which is displayed in Fig. 1.17.

This combined accelerating mechanism has great potential to realize tens-of-GeV-level table-tap size accelerators and is attracting more and more attentions. There also remain many open problems. In [99], it is mentioned that trapped ions can be confined in the bubble only with CP lasers. The reason is not yet understood. It definitely calls for further researches.

1.4 Intense High-Order Harmonics and Attosecond Pulses

High-order harmonics and attosecond pulses (APs) are able to resolve electron motion in atoms or molecules, thus play important roles in studying sub-femtosecond, attosecond processes. The main approach to produce such light field is using moderate intensity (10^{14}–10^{16} W/cm^2) laser to interact with atoms (gas). Under the laser field, electrons in atoms escape through field ionization; then are accelerated by the laser and finally combine with nucleus again by radiating high harmonics. The abundant harmonics exhibit as an attosecond chain in the time domain [102–104]. This approach has made great progress and shows good prospects. Meanwhile, we should also notice that the generated light field is relatively low as the intensity of the driving laser is limited by atom ionization threshold. Besides, the energy efficiency is not so high, only about 10^{-7}–10^{-6}.

In most applications, single isolated attosecond pulse is essential. It is quite difficult to isolated one AP from a chain. One way is using single-cycle laser pulse with stable carrier envelope phase (CEP). This can only be realized by very few labs in the world. Another way is to use polarization gate technique, which requires extraordinarily accurate control on light path.

Bulanov et al. [105] discovered that when a relativistic laser pulse interacts with a plasma target of solid density, the plasma boundary oscillates violently and reflects the incident pulse simultaneously. High harmonics are induced in the reflected pulse due to the Doppler effect. This mechanism was further developed [106–108] and now referred as the "relativistic oscillating mirror (ROM)" model. Since it is a plasma approach, laser intensity is no longer limited by ionization threshold, thus can produce ultra-intense high harmonics and APs.

1.4.1 Relativistic Oscillating Mirror Model

The ponderomitve force of LP pulses includes a high-frequency component, which oscillates at twice the laser frequency. When it is incident on a target, it can drive strong oscillation of the plasma boundary. The oscillating velocity can be very close to the light speed as long as the laser pulse is relativistic. From the electron distribution in Fig. 1.18, one notices that the boundary oscillating period is half of the laser period.

Reflected by such a "mirror", the laser field is modified to

$$E_r[t + X(t)/c] = -E_i[t - X(t)/c]. \tag{1.65}$$

where E_r and E_i is the reflected and incident laser field respectively. $X(t)$ is instantaneous position of the oscillating boundary. Early theoretical researches [105, 106] considered that harmonics originate from the modified laser field caused by the variation of boundary position $X(t)$. Later on Gordienko et al. [107] analyzed the spectra of reflected laser fields. They found that in the highly relativistic regime, the harmonic intensity scales with $I_\omega \sim \omega^{-5/2}$, as shown in Fig. 1.19. According to the Doppler effect, the frequency of reflected laser is

$$\frac{\omega_r}{\omega_i} = \frac{1+\beta}{1-\beta} \approx 4\gamma^2. \tag{1.66}$$

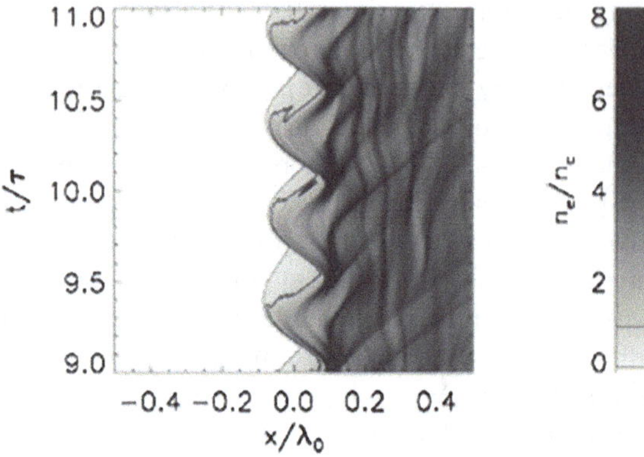

Fig. 1.18 The evolvement of the electron density [106]

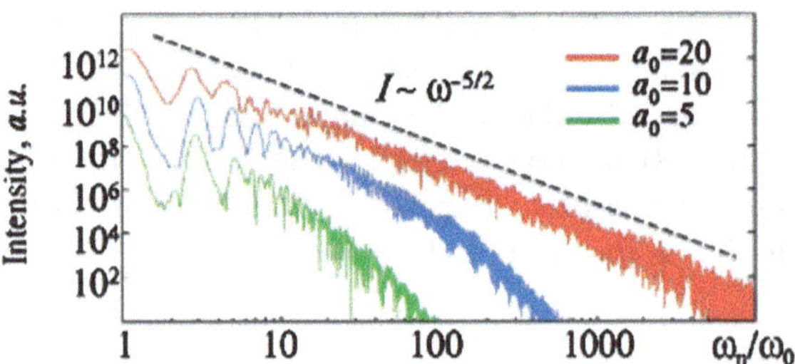

Fig. 1.19 Spectra of the reflected laser fields [107]

here β is the boundary velocity normalized by the light speed. The cut-off frequency is then about $\omega_{\max} \approx 4\gamma_{\max}^2$. The spectrum is also not continuous. This is due to multiple oscillations of the boundary.

Baeva et al. [108] further developed the model. They believed that one oscillation leads to a sudden change in the velocity domain and hence a spike in the γ^2 domain, which is displayed in Fig. 1.20. According to Eq. (1.66), a bunch of harmonics are induced. Through more religious calculation, they gave a scaling law of $I_\omega \sim \omega^{-8/3}$, a little steeper than previously predicted. The cut-off frequency is also renewed to $\omega_{\max} \approx 8^{1/2}\gamma_{\max}^3$. One can find these results in Fig. 1.21.

Though the theories are one dimensional, they agree with simulation results very well. Furthermore, they explain perfectly the experimental results [110–112].

1.4.2 Intense Attosecond Pulses

One of the most important applications of high harmonics should be generating attosecond pulses (APs) [113]. In ROM mechanism, LP laser pulse will drive two oscillations in one laser period. Each oscillation generates a γ^2 spike, inducing harmonics of orders up to several thousand. In the time domain, an AP appears for

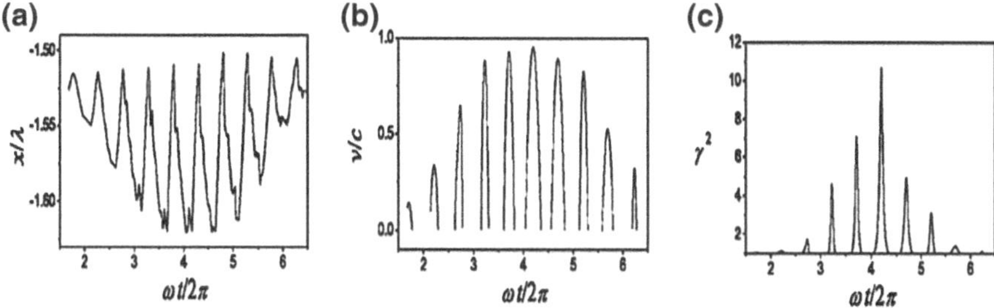

Fig. 1.20 The position, velocity and the square of the relativistic factor of the oscillating boundary [108]

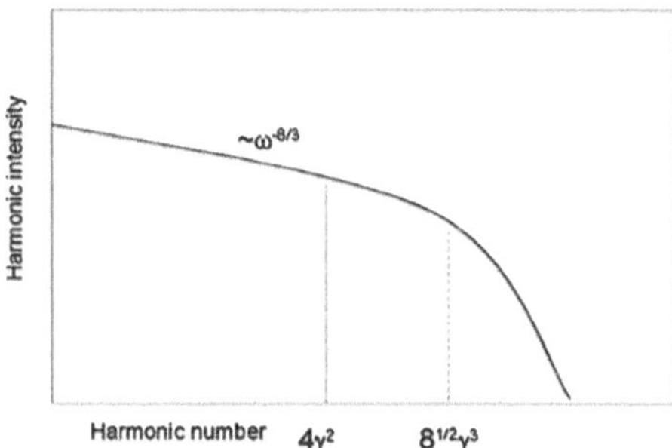

Fig. 1.21 Spectrum of the reflected laser field: the scaling law and the cut-off frequency [109]

one oscillation. Normally the pulse duration is more than one laser period, which leads to an AP train, while most related applications require a single isolated pulse. The ROM mechanism thus faces the same difficulty as in the laser-atom scheme-isolating AP from a train. Naumova et al. [114, 115] proposed to use a laser pulse focused in a volume of a few cubic wavelengths. The reflecting surface is strongly modified due to the tightly focused laser field, i.e., the reflecting direction changes at every moment. In such a manner, APs are reflected to different directions and isolated, as shown in Fig. 1.22c. This showed the possibility of generating single AP in ROM for the first time. Later on, other methods were also proposed, such as using precise frequency filtering, polarization gating technique etc. [114–118].

1.4.3 Multi-dimensional Effects

Multi-dimensional effects are very important in high harmonic and AP generation. In last section, APs are split due to these effects. On the one hand, multi-dimensional instabilities such as Rayleigh-Taylor-like and Weibel-like instabilities will deform the target surface and modify the density profile. This may reduce the spatial coherence of harmonics. On the other hand, when a laser pulse impinges on

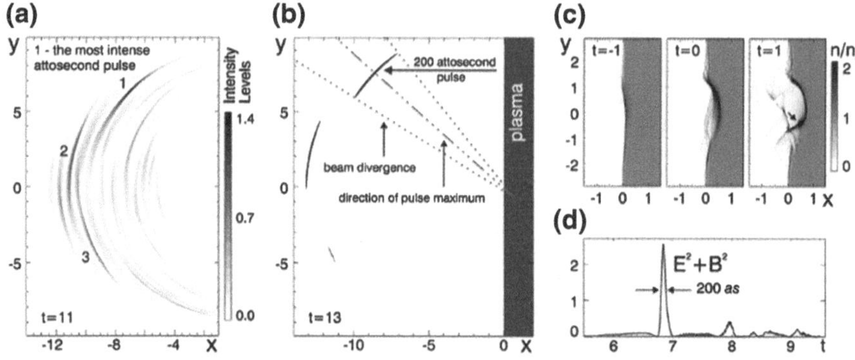

Fig. 1.22 Pulse separation by using a cubic-wavelength laser pulse [114]

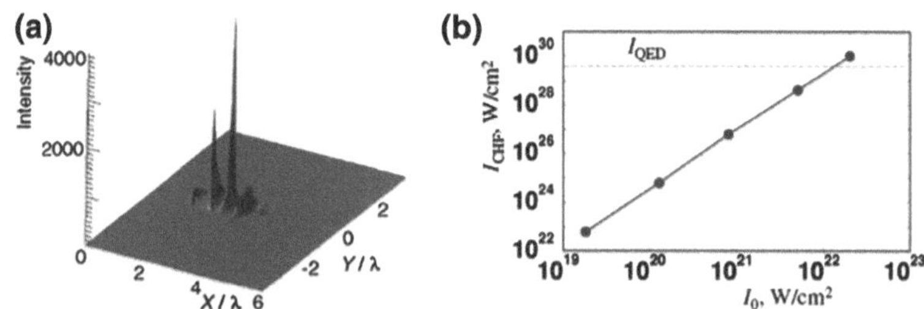

Fig. 1.23 Laser intensity distribution after being coherently focused (a) and the scaling law between focused and incident intensities (b) [119]

a spherically recessed target, light intensity can be greatly enhanced due to coherent harmonic focusing (CHF) of the reflecting surface. The focused intensity may even reach the Schwinger limit [119].

The decaying law of harmonic spectrum is crucial for coherent focusing. For incoherent harmonics, the focal intensity is the sum of all harmonics. Since the minimum size of the focal spot is inversely proportional to the harmonic frequency, i.e., $\sim 1/\omega$, the peak focal intensity is boosted if the harmonic decaying law is slower than $1/\omega^2$. Otherwise the focused intensity of a harmonic will decrease with higher order. While the harmonics are coherent, it turns out that a boosted focusing can happen only if the decaying law is slower than $1/\omega^4$. As been introduced in Sect. 1.4.1, harmonics from ROM mechanism are coherent and with a scaling law of $I_\omega \sim \omega^{-8/3}$, which apparently is perfect for CHF. The peak intensity of the reflected laser pulse from an inner concave target surface is dramatically enhanced. Figure 1.23 shows that the peak intensity from coherent focusing is higher than the linear superposition intensity by at least one magnitude. Figure 1.23b gives the scaling law between focused and incident intensities, which indicates that 10^{22} W/cm^2 laser is strong enough to produce light intensity close to the Schwinger limit.

References

1. D. Strickland, G. Mourou, Compression of amplified chirped optical pulses. Opt. Commun. **56**, 219–223 (1985)
2. M.D. Perry, G.A. Mourou, Terawatt to petawatt subpicosecond laser. Science **264**, 917–924 (1994)
3. G. Mourou, C.P. Batty, M.D. Perry, Ultrashort intensity lasers: physics of the extreme on a tabtop. Phys. Today **51**, 22–24 (1998)
4. W. Liu, A brief introduction of plasma physics, pp. 1, (2002) (in chinese)
5. T. Tajima, G. Mourou, Zettawatt-exawatt lasers and their applications in ultrastrong field physics. Phys. Rev. ST Accel. Beams **5**, 031301 (2002)
6. D. Umstadter, Relativistic laser-plasma interactions. J. Phys. D Appl. Phys. **36**, R151–R165 (2003)
7. Z. Sheng, Relativistic laser-plasma interaction-A lecture in USTC (2006) (in chinese)
8. T. Tajima, J.M. Dawson, Laser electron accelerator. Phys. Rev. Lett. **43**, 267–270 (1979)
9. J.M. Dawson, Nonlinear electron oscillations in a cold plasma. Phys. Rev. **133**, 383–387 (1959)
10. E. Esarey, A. Ting, P. Sprangle, D. Umstadter, X. Liu, Nonlinear analysis of relativistic harmonic generation by intense lasers in plasmas. IEEE Trans. Plasma Sci. **21**, 95–104 (1993)
11. E. Esarey, P. Sprangle, J. Krall, A. Ting, Overview of plasma-based accelerator concepts. IEEE Trans. Plasma Sci. **24**, 252 (1996)
12. J.D. Jackson, in *Classical Electrodynamics* (translated by Peiyu Zhu), vol. 1, (Wiley, Newyork, 1975) p. 241
13. J.D. Jackson, in *Classical Electrodynamics* (translated by Peiyu Zhu), vol. 2, (Wiley, Newyork, 1975) pp. 125–128
14. C.S. Lai, Phys. Rev. Lett. **36**, 966–968 (1976)
15. W. Yu, M.Y. Yu, Z.M. Sheng, J. Zhang, Model for fast electrons in ultrashort-pulse laser interaction with solid targets. Phys. Rev. E **58**, 2456–2460 (1998)
16. B. Shen, J. Meyer-ter-Vehn, High-density ($>10^{23}/cm^3$) relativistic electron plasma confined between two laser pulses in a thin foil. Phys. Plasmas **8**, 1003–1010 (2001)
17. R.E.W. Pfund, R. Lichters, J. Meyer-ter-Vehn, in *Super Strong Fields in Plasmas*, ed. by M. Lontano et al., AIP Conference Proceedings, vol. 426 (American Institute of Physics, Melville, 1998), p. 141
18. S. Eliezer, *The Interaction of High-Power Lasers With Plasmas* (Institute of Physics Publishing, London, 2002)
19. M. Tabak, J. Hammer, M.E. Glinsky, W.L. Kruer, S.C. Wilks, J. Woodworth, E.M. Campbell, M.D. Perry, Ignition and high gain with ultrapowerful lasers. Phys. Plasmas **1**, 1626–1634 (1994)
20. M. Tabak, D.S. Clark, S.P. Hatchett, M.H. Key, B.F. Lasinski, R.A. Snavely, S.C. Wilks, R.P.J. Town, R. Stephens, E.M. Campbell, R. Kodama, K. Mima, K.A, Tanaka, Review of progress in fast ignition. Phys. Plasmas **12**, 057305 (2005)
21. J. Meyer-ter-Vehn, Fast ignition of ICF targets: an overview. Plasma Phys. Contr. Fusion **43**, A113–A118 (2001)
22. A. Pukhov, J. Meyer-ter-Vehn, Relativistic magnetic self-channeling of light in near-critical plasma: three-dimensional particle-in-cell simulation. Phys. Rev. Lett. **76**, 3975–3978 (1996)
23. A. Pukhov, J. Meyer-ter-Vehn, Laser hole boring into overdense plasma and relativistic electron currents for fast ignition of ICF targets. Phys. Rev. Lett. **79**, 2686–2689 (1997)
24. A. Pukhov, Z. M. Sheng, J. Meyer-ter-Vehn, Particle acceleration in relativistic laser channels. Phys. Plasmas **6**, 2847 (1999)

25. Z.-M. Sheng, K. Mima, J. Zhang, J. Meyer-ter-Vehn, Efficient acceleration of electrons with counterpropagating intense laser pulses in vacuum and underdense plasma. Phys. Rev. E **69**, 016407 (2004)
26. M.-C. Firpo, A.F. Lifschitz, E. Lefebvre, C. Deutsch, Early out-of-equilibrium beam-plasma evolution. Phys. Rev. Lett. **96**, 115004 (2006)
27. J.J. Honrubia, J. Meyer-ter-Vehn, Three-dimensional fast electron transport for ignition-scale inertial fusion capsules. Nucl. Fusion **46**, L25–L28 (2006)
28. S. Atzeni, J. Meyer-ter-Vehn, *The Physics of Inertial Fusion* (Oxford University, New York, 2004)
29. R. Kodama, P.A. Norreys, K. Mima, A.E. Dangor, R.G. Evans, H. Fujita, Y. Kitagawa, K. Krushelnick, T. Miyakoshi, N. Miyanaga, T. Norimatsu, S.J. Rose, T. Shozaki, K. Shigemori, A. Sunahara, M. Tampo, K.A. Tanaka, Y. Toyama, T. Yamanaka, M. Zepf, Nature **412**, 798–802 (2001)
30. R. Kodama, H. Shiraga, K. Shigemori, Y. Toyama, S. Fujioka, H. Azechi, H. Fujita, H. Habara, T. Hall, Y. Izawa, T. Jitsuno, Y. Kitagawa, K.M. Krushelnick, K.L. Lancaster, K. Mima, K. Nagai, M. Nakai, H. Nishimura, T. Norimatsu, P.A. Norreys, S. Sakabe, K.A. Tanaka, A. Youssef, M. Zepf, T. Yamanaka, Nature **418**, 933–934 (2002)
31. A.L. Lei, K.A. Tanaka, R. Kodama, G.R. Kumar, K. Nagai, T. Norimatsu, T. Yabuuchi, K. Mima, Optimum hot electron production with low-density foams for laser fusion by fast ignition. Phys. Rev. Lett. **96**, 255006 (2006)
32. J.M. Dawson, One-dimensional plasma model. Phys. Fluids **5**, 445 (1962)
33. W. Yu, H. Xu, F. He, M.Y. Yu, S. Ishiguro, J. Zhang, A.Y. Wong, Direct acceleration of solid-density plasma bunch by ultraintense laser. Phys. Rev. E **72**, 046401 (2005)
34. G.J. Pert, The analytic theory of linear resonant absorption. Plasma Phys. **20**, 175–188 (1978)
35. F. Brunel, Not-so-resonant, resonant absorption. Phys. Rev. Lett. **59**, 52–55 (1987)
36. S.C. Wilks, W.L. Kruer, M. Tabak, A.B. Langdon, Absorption of ultra-intense laser pulses. Phys. Rev. Lett. **69**, 1383–1386 (1992)
37. J. Denavit, Absorption of high-intensity subpicosecond lasers on solid density targets. Phys. Rev. Lett. **69**, 3052–3055 (1992)
38. J.T. Mendonca, Threshold for electron heating by two electronmagnetic wave. Phys. Rev. A **28**, 3592 (1983)
39. J.M. Rax, Compton harmonic resonance, stochastic instabilities quasilinear diffusion and collisionless damping with ultrahigh intensity laser waves. Phys. Fluids **4**, 3962 (1992)
40. Z.M. Sheng, K. Mima, Y. Sentoku, M.S. Jovanovic, T. Taguchi, J. Zhang, J. Meyer-ter-Vehn, Stochastic heating and acceleration of electrons in colliding laser fields in plasma. Phys. Rev. Lett. **88**, 055004 (2002)
41. A. Pukhov, Three-dimensional electromagnetic relativistic particle-in-cell code VLPL (Virtual Laser Plasma lab). J. Phys. Plasma **61**, 425–433 (1999)
42. A. Pukhov, J. Meyer-ter-Vehn, Relativistic laser-plasma interaction by multi-dimensional particle-in-cell simulations. Phys. Plasmas **1998**, 5 (1880)
43. S. Atzeni, J. Meyer-ter-Vehn, *The Physics of Inertial Fusion* (translated by Baifei Shen), (Oxford University Press, Newyork, 2004) pp. 394
44. S.V. Blanov, F. Pegoraro, A.M. Pukhov, Two-dimensional regimes of self-focusing, wake field generation, and induced focusing of a short intense laser pulse in an underdense plasma. Phys. Rev. Lett. **74**, 710–713 (1995)
45. E. Esarey, B. Hafizi, R. Hubbard, A. Ting, Trapping and acceleration in self-modulational laser wakefields. Phys. Rev. Lett. **80**, 5552–5555 (1998)
46. V. Malka, S. Fritzler, E. Lefebvre, M.M. Aleonard, F. Burgy, J.P. Chambaret, J.F. Chemin, K. Krushelnick, G. Malka, S.P.D. Mangles, Z. Najmudin, M. Pittman, J.P. Rosseau, J.N. Scheurer, B. Walton, A.E. Dangor, Electron acceleration by a wake field forced by an intense ultrashort laser pulse. Science **298**, 1596–1600 (2002)
47. S.P.D. Mangles, C.D. Murphy, Z. Najmudin, A.G.R. Thomas, J.L. Collier, A.E. Dangor, E.J. Divall, P.S. Foster, J.G. Gallacher, C.J. Hooker, D.A. Jaroszynski, A.J. Langley, W.B.

Mon, P.A. Norreys, F.S. Tsung, R. Viskup, B.R. Walton, K. Krushelnick, Nature **431**, 535–538 (2004)

48. C.G.R. Geddes, Cs. Toth, J. van Tilborg, E. Esarey, C.B. Schroeder, D. Bruhwiler, C. Nieter, J. Cary, W.P. Leemans, High-quality electron beams from a laser wakefield accelerator using plasma-channel guiding. Nature **431**, 538–541 (2004)
49. J. Faure, Y. Glinec, A. Pukhov, S. Kiselev, S. Gordienko, E. Lefebvre, J.P. Rousseau, F. Burgy, V. Malka, A laser-plasma accelerator producing monoenergetic electron beams. Nature **431**, 541–544 (2004)
50. N. Ishii, L. Turi, V.S. Yakovlev, T. Fuji, F. Krausz, A. Baltuska, R. Butkus, G. Veitas, V. Smilgevicius, R. Danielius, A. Piskarskas, Multimillijoule chirped parametric amplification of few-cycle pulses. Opt. Lett. **30**, 567 (2005)
51. B. Hidding, K.U. Amthor, B. Liesfeld, H. Schwoerer, S. Karsch, M. Geissler, L. Veisz, K. Schmid, J.G. Gallacher, S.P. Jamison, D. Jaroszynski, G. Pretzler, R. Sauerbrey, Generation of quasimonoenergetic electron bunches with 80 fs laser pulses. Phys. Rev. Lett. **96**, 105004 (2006)
52. A. Pukhov, J. Meyer-ter-Vehn, Laser wake field acceleration: the highly non-linear broken-wave regime. Appl. Phys. B **74**, 355–361 (2002)
53. I. Kostyukov, A. Pukhov, S. Kiselev, Phenomenological theory of laser-plasma interaction in "bubble" regime. Phys. Plasmas **11**, 5256 (2004)
54. W.P. Leemans, B. Nagler, A.J. Gonsalves, Cs. Tóth, K. Nakamura, C.G.R. Geddes, E. Esarey, C.B. Schroeder, S.M. Hooker, GeV electron beams from a centimetre-scale accelerator. Nat. Phys. **2**, 696–699 (2006)
55. C. Joshi, Plasma accelerators. Sci. Am. **294**, 41–47 (2006)
56. J. Faure, C. Rechantin, A. Norlin, A. Lifschitz, Y. Glinec, V. Malka, Controlled injection and acceleration of electrons in plasma wakefields by colliding laser pulses. Nature **444**, 737–739 (2006)
57. E. Esarey, R.F. Hubbard, W.P. Leemans, A. Ting, P. Sprangle, Electron injection into plasma wakefields by colliding laser pulses. Phys. Rev. Lett. **79**, 2682–2685 (1997)
58. B. Shen, Y. Li, K. Nemeth, H. Shang, Y. Chae, R. Soliday, R. Crowell, E. Frank, W. Gropp, J. Cary, Electron injection by a nanowire in the bubble regime. Phys. Plasmas **14**, 053115 (2007)
59. A. Pukhov, Three-dimensional simulations of ion acceleration from a foil irradiated by a short-pulse laser. Phys. Rev. Lett. **86**, 3562–3565 (2001)
60. S.C. Wilks, A.B. Langdon, T.E. Cowan, M. Roth, M. Singh, S. Hatchett, M.H. Key, D. Pennington, A. MacKinnon, R.A. Snavely, Energetic proton generation in ultra-intense laser–solid interactions. Phys. Plasmas **8**, 542–549 (2001)
61. L. Romagnani, J. Fuchs, M. Borghesi, P. Antici, P. Audebert, F. Ceccherini, T. Cowan, T. Grismayer, S. Kar, A. Macchi, P. Mora, G. Pretzier, A. Schiavi, T. Toncian, O. Willi, Dynamics of electric fields driving the laser acceleration of multi-MeV protons. Phys. Rev. Lett. **95**, 195001 (2005)
62. S.P. Hatchett, C.G. Brown, T.E. Cowan, E.A. Henry, J.S. Johnson, M.H. Key, J.A. Koch, A.B. Langdon, B.F. Lasinski, R.W. Lee, A.J. Mackinnon, D.M. Pennington, M.D. Perry, T.W. Phillips, M. Roth, T.C. Sangster, M.S. Singh, R.A. Snavely, M.A. Stoyer, S.C. Wilks, K. Yasuike, Electron, photon, and ion beams from the relativistic interaction of Petawatt laser pulses with solid targets. Phys. Plasmas **7**, 2076–2082 (2000)
63. R. Snavely, M. Key, S. Hatchett, T.E. Cowan, M. Roth, T.W. Phillips, M.A. Stoyer, E.A. Henry, T.C. Sangster, M.S. Singh, S.C. Wilks, A. MacKinnon, A.A. Offenberger, D.M. Pennington, K. Yasuike, A.B. Langdon, B.F. Lasinski, M.D. Perry, E.M. Campbell, Intense high-energy proton beams from Petawatt-laser irradiation of solids. Phys. Rev. Lett. **85**, 2945–2948 (2000)
64. M. Roth, A. Blazevic, M. Geissel, T. Schlegel, T.E. Cowan, M. Allen, J.C. Gauthier, P. Audebert, J. Fuchs, J. Meyer-ter-Vehn, M. Hegelich, S. Karsch, A. Pukhov, Energetic ions generated by laser pulses: a detailed study on target properties. Phys. Rev. ST Accel. Beams **5**, 061301 (2002)

65. M. Hegelich, S. Karsch, G. Pretzler, D. Habs, K. Witte, W. Guenther, M. Allen, A. Blazevic, J. Fuchs, J.C. Gauthier, M. Geissel, P. Audebert, T. Cowan, M. Roth, MeV ion jets from short-pulse-laser interaction with thin foils. Phys. Rev. Lett. **89**, 085002 (2002)
66. B.M. Hegelich, *Ultra-high intensity laser acceleration of ions to MeV/nucleon energies* (Particle accelerator conference, Albuquerque, 2007), pp. 25–29
67. S.C. Wilks, Simulations of ultraintense laser-plasma interactions. Phys. Fluids B **5**, 2603–2608 (1993)
68. P. Mora, Plasma expansion into a vacuum. Phys. Rev. Lett. **90**, 185002 (2003)
69. TZh Esirkepov, S.V. Bulanov, K. Nishihara, T. Tajima, F. Pegoraro, V.S. Khoroshkov, K. Mima, H. Daido, Y. Kato, Y. Kitagawa, K. Nagai, S. Sakabe, Proposed double-layer target for the generation of high-quality laser-accelerated ion beams. Phys. Rev. Lett. **89**, 175003 (2002)
70. B.M. Hegelich, B.J. Albright, J. Cobble, K. Flippo, S. Letzring, M. Paffett, H. Ruhl, J. Schreiber, R.K. Schulze, J.C. Fernandez, Laser acceleration of quasi-monoenergetic MeV ion beams. Nature **439**, 441 (2006)
71. S.V. Bulanov, V.S. Khoroshkov, Feasibility of using laser ion accelerators in proton therapy. Plasma Phys. Rep. **28**, 453–456 (2002)
72. H. Schwoerer, S. Pfotenhauerl, O. Jaeckel, K.-U. Amthor, B. Liesfeld, W. Ziegler, R. Sauerbrey, K.W.D. Ledingham, T. Esirkepov, Laser-plasma acceleration of quasi-monoenergetic protons from microstructured targets. Nature **439**, 445 (2006)
73. T. Toncian, M. Borghesi, J. Fuchs, E. d'Humieres, P. Antici, P. Audebert, E. Brambrink, C.A. Cecchetti, A. Pipahl, L. Romagnani, O. Willi, Ultrafast-driven microlens to focus and energy-select Mega-Electron-Volt protons. Science **312**, 410–413 (2006)
74. Y. Nodera, S. Kawata, N. Onuma, J. Limpouch, O. Klimo, T. Kikuchi, Improvement of energy-conversion efficiency from laser to proton beam in a laser-foil interaction. Phys. Rev. E **78**, 046401 (2008)
75. F. Wang, B. Shen, X. Zhang, Z. Jin, M. Wen, L. Ji, W. Wang, J. Xu, M. Yu, J. Cary, High-energy monoenergetic proton bunch from laser interaction with a complex target. Phys. Plasmas **16**, 093112 (2009)
76. L. Yin, B.J. Albright, B.M. Hegelich, K.J. Browers, K.A. Flippo, T.J.T. Kwan, J.C. Fernandez, Monoenergetic and GeV ion acceleration from the laser breakout afterburner using ultrathin targets. Phys. Plasmas **14**, 056706 (2007)
77. J. Zheng, Theory of plasma physics-A lecture in USTC, pp. 110 ("in chinese")
78. D.A. Tidman, N.A. Krall, *Shock Waves in Collisionless Plasmas* (Wiley-Interscience, New York, 1971), pp. 99–112
79. Y Chen, An introduction of nonlinear plasma physics-A lecture in USTC ("in chinese")
80. D.W. Forslund, C.R. Shonk, Formation and structure of electrostatic collisionless shocks. Phys. Rev. Lett. **25**, 1699–1702 (1970)
81. S.C. Wilks, W.L. Kruer, Absorption of ultrashort, ultra-intense laser light by solids and overdense plasmas. IEEE J. Quantum Electron. **33**, 1954–1968 (1997)
82. J. Denavit, Absorption of high-intensity subpicosecond lasers on solid density targets. Phys. Rev. Lett. **69**, 3052–3055 (1992)
83. L.O. Silva, M. Marti, J.R. Davis, R.A. Fonseca, Proton shock acceleration in laser-plasma interactions. Phys. Rev. Lett. **92**, 015002 (2004)
84. H. Habara, K.L. Lancaster, S. Karsch, C.D. Murphy, P.A. Norreys, R.G. Evans, M. Borghesi, L. Romagnani, M. Zepf, T. Norimatsu, Y. Toyama, R. Kodama, J.A. King, R. Snavely, K. Akli, B. Zhang, R. Freeman, S. Hatchett, A.J. MacKinnon, P. Patel, M.H. Key, C. Stoeckl, R.B. Stephens, R.A. Fonseca, L.O. Silva, Ion acceleration from the shock front induced by hole boring in ultraintense laser-plasma interactions. Phys. Rev. E **70**, 046414 (2004)
85. M. Chen, Z. Sheng, Q. Dong, M. He, Y. Li, M.A. Bari, J. Zhang, Collisionless electrostatic shock generation and ion acceleration by ultraintense laser pulses in overdense plasmas. Phys. Plasmas **14**, 053102 (2007)

86. A. Macchi, F. Cattani, T.V. Liseykina, F. Cornolti, Laser acceleration of ion bunches at the front surface of overdense plasmas. Phys. Rev. Lett. **94,** 165003 (2005)
87. X. Zhang, B. Shen, X. Li, Z. Jin, F. Wang, M. Wen, Efficient GeV ion generation by ultraintense circularly polarized laser pulse. Phys. Plasmas **14**, 123108 (2007)
88. X. Zhang, B. Shen, Z. Jin, F. Wang, L. Ji, Generation of plasma intrinsic oscillation at the front surface of a target irradiated by a circularly polarized laser pulse. Phys. Plasmas **16**, 033102 (2009)
89. X. Zhang, B. Shen, X. Li, Z. Jin, F. Wang, Multi-staged acceleration of ions by circularly polarized laser pulse: monoenergetic ion beam generation. Phys. Plasmas **14**, 073101 (2007)
90. B. Shen, Z. Xu, Transparency of an overdense plasma layer. Phys. Rev. E **64**, 056406 (2001)
91. X.Q. Yan, C. Lin, Z.M. Sheng, Z.Y. Guo, B.C. Liu, Y.R. Lu, J.X. Fang, J.E. Chen, Generating high-current monoenergetic proton beams by a circularly polarized laser pulse in the phase-stable acceleration regime. Phys. Rev. Lett. **100**, 135003 (2008)
92. T. Esirkepov, M. Borghesi, S.V. Bulanov, G. Mourou, T. Tajima, Highly efficient relativistic-ion generation in the laser-piston regime. Phys. Rev. Lett. **92**, 175001 (2004)
93. A. Henig, S. Steinke, M. Schnürer, T. Sokollik, R. Hörlein, D. Kiefer, D. Jung, J. Schreiber, B.M. Hegelich, X.Q. Yan, J. Meyer-ter-Vehn, T. Tajima, P.V. Nickles, W. Sandner, D. Habs, Radiation-pressure acceleration of ion beams driven by circularly polarized laser pulses. Phys. Rev. Lett. **103**, 245003 (2009)
94. M. Chen, A. Pukhov, Z.M. Sheng, X.Q. Yan, Laser mode effects on the ion acceleration during circularly polarized laser pulse interaction with foil targets. Phys. Plasmas **15**, 113103 (2008)
95. X.Q. Yan, H.C. Wu, Z.M. Sheng, J.E. Chen, J. Meyer-ter-Vehn, Self-organizing GeV, nanocoulomb, collimated proton beam from laser foil interaction at 7e21 W/cm^2. Phys. Rev. Lett. **103**, 135001 (2009)
96. B. Qiao, M. Zepf, M. Borghesi, M. Geissler, Stable GeV ion-beam acceleration from thin foils by circularly polarized laser pulses. Phys. Rev. Lett. **102**, 145002 (2009)
97. T.P. Yu, A. Pukhov, G. Shvets, M. Chen, Stable laser-driven proton beam acceleration from a two-ion-species ultrathin foil. Phys. Rev. Lett. **105**, 065002 (2010)
98. B. Shen. Y. Li, M.Y. Yu, J. Cary, Bubble regime for ion acceleration in a laser-driven plasma. Phys. Rev. E **76**, 055402(R) (2007)
99. B. Shen, X. Zhang, Z. Sheng, M.Y. Yu, J. Cary, High-quality monoenergetic proton generation by sequential radiation pressure and bubble acceleration. Phys. Rev. ST Accel. Beams **12**, 121301 (2009)
100. X. Zhang, B. Shen, L. Ji, F. Wang, M. Wen, W. Wang, J. Xu, Y. Yu, Ultrahigh energy proton generation in sequential radiation pressure and bubble regime. Phys. Plasmas **17**, 123102 (2010)
101. L.L. Yu, H. Xu, W.M. Wang, Z.M. Sheng, B.F. Shen, W. Yu, J. Zhang, Generation of tens of GeV quasi-monoenergetic proton beams from a moving double layer formed by ultraintense lasers at intensity 10^{21}–10^{23} Wcm^{-2}. New J. Phys. **12**, 045021 (2010)
102. G. Sansone, E. Benedetti, F. Calegari, C. Vozzi, L. Avaldi, R. Flammini, L. Poletto, P. Villoresi, C. Altucci, R. Velotta, S. Stagira, S. De Silvestri, M. Nisoli, Isolated single-cycle attosecond pulse. Science **314**, 443–446 (2006)
103. E. Goulielmakis, M. Schultze, M. Hofstetter, V.S. Yakovlev, J. Gagnon, M. Uiberacker, A.L. Aquila, E.M. Gullikson, D.T. Attwood, R. Kienberger, F. Krausz, U. Kleineberg, Single-cycle nonlinear optics. Science **320**, 1614–1617 (2008)
104. X. Feng, S. Gilbertson, H. Mashiko, H. Wang, S.D. Khan, M. Chini, Y. Wu, K. Zhao, Z. Chang, Generation of isolated attosecond pulses with 20–28 femtosecond lasers. Phys. Rev. Lett. **103**, 183901 (2009)
105. S.V. Bulanov, N.M. Naumova, F. Pegoraro, Interaction of an ultrashort, relativistically strong laser pulse with an overdense plasma. Phys. Plasmas **1**, 745–757 (1994)
106. R. Lichters, J. Meyer-ter-Vehn, A. Pukhov, Short-pulse laser harmonics from oscillating plasma surfaces driven at relativistic intensity. Phys. Plasmas **3**, 3425–3437 (1996)

107. S. Gordienko, A. Pukhov, O. Shorokhov, T. Baeva, Relativistic Doppler effect: universal spectra and zeptosecond pulses. Phys. Rev. Lett. **93**, 115002 (2004)
108. T. Baeva, S. Gordienko, A. Pukhov, Theory of high-order harmonic generation in relativistic laser interaction with overdense plasma. Phys. Rev. E **74**, 046404 (2006)
109. U. Teubner, P. Gibbon, High-order harmonics from laser-irradiated plasma surfaces. Rev. Mod. Phys. **81**, 445–479 (2009)
110. B. Dromey, S. Kar, C. Bellei, D.C. Carroll, R.J. Clarke, J.S. Green, S. Kneip, K. Markey, S.R. Nagel, P.T. Simpson, L. Willingale, P. Mckenna, D. Neely, Z. Najmudin, K. Krushelnick, P.A. Norreys, M. Zepf, Bright multi-keV harmonic generation from relativistically oscillating plasma surfaces. Phys. Rev. Lett. **99**, 085001 (2007)
111. B. Dromey, M. Zepf, A. Gopal, K. Lancaster, M.S. Wei, K. Krushelnick, M. Tatarakis, N. Vakakis, S. Moustaizis, R. Kodama, M. Tampo, C. Stoeckl, R. Clarke, H. Habara, D. Neely, S. Karsch, P. Norreys, High harmonic generation in the relativistic limit. Nat. Phys. **2**, 456–459 (2006)
112. B. Dromey, D. Adams, R. Hörlein, Y. Nomura, S.G. Rykovanov, D.C. Carroll, P.S. Foster, S. Kar, K. Markey, P. McKenna, D. Neely, M. Geissler, G.D. Tsakiris, M. Zepf, Diffraction-limited performance and focusing of high harmonics from relativistic plasmas. Nat. Phys. **5**, 146–152 (2009)
113. L. Plaja, L. Roso, K. Rzazewski, M. Lewenstein, Generation of attosecond pulse trains during the reflection of a very intense laser on a solid surface. J. Opt. Soc. Am. B **15**, 1904–1911 (1998)
114. N.M. Naumova, J.A. Nees, I.V. Sokolov, B. Hou, G.A. Mourou, Relativistic generation of isolated attosecond pulse in a λ^3 focal volume. Phys. Rev. Lett. **92**, 063902 (2004)
115. N.M. Naumova, J.A. Nees, G.A. Mourou, Relativistic attosecond physics. Phys. Plasmas **12**, 056707 (2005)
116. T. Baeva, S. Gordienko, A. Pukhov, Relativistic plasma control for single attosecond x-ray burst generation. Phys. Rev. E **74**, 065401(R) (2006)
117. L. Liu, C. Xia, J. Liu, W. Wang, Y. Cai, C. Wang, R. Li, Z. Xu, Control of single attosecond pulse generation from the reflection of a synthesized relativistic laser pulse on a solid surface. Phys. Plasmas **15**, 103107 (2008)
118. S.G. Rykovanov, M. Geissler, J. Meyer-ter-Vehn, G.D. Tsakiris, Intense single attosecond pulses from surface harmonics using the polarization gating technique. New J. Phys. **10**, 025025 (2008)
119. S. Gordienko, A. Pukhov, O. Shorokhov, T. Baeva, Coherent focusing of high harmonics: a new way towards the extreme intensities. Phys. Rev. Lett. **94**(103903), 1 (2005)
120. YuM Mikhailova, V.T. Platonenko, S.G. Rykovanov, Generation of an attosecond x-ray pulse in a thin film irradiated by an ultrashort ultrarelativistic laser pulse. JETP Lett. **81**, 571–574 (2005)
121. A.S. Pirozhkov, S.V. Bulanov, T.Z. Esirkepov, M. Mori, A. Sagisaka, H. Daido, Attosecond pulse generation in the relativistic regime of the laser-foil interaction: the sliding mirror model. Phys. Plasmas **13**, 013107 (2006)
122. D. an der Brügge, A. Pukhov, Enhanced relativistic harmonics by electron nanobunching. Phys. Plasmas **17**, 033110, (2010)

Chapter 2
Ion Acceleration I: Efficient Heavy Ion Acceleration by ESA

2.1 Introduction

Monoenergetic ion beams play essential roles in many important applications such as fast ignition fusion, cancer therapy, and other applications [1]. The new accelerating method, laser ion acceleration, attracts more and more attention nowadays since the acceleration gradient is at least four magnitudes higher than that of conventional methods. A number of novel mechanisms have been proposed for accelerating protons or heavier ions to high energies using the recently available high-intensity short-pulse lasers. With target-normal sheath acceleration, protons up to 58 MeV have been obtained with 3×10^{20} W/cm^2 lasers [2]. With microstructured targets, quasi-monoenergetic proton or ion bunches with energy spread of about 20 % have been produced [3, 4]. Other promising schemes include shock acceleration at the front of the foil [5–9], direct laser-pressure acceleration [10, 11], and acceleration by plasma wake [12].

Despite these improvements, acceleration of heavy ions by laser plasma interaction is still a challenge [13] because of certain physical limits. Heavy ions cannot be accelerated as efficiently as protons because of their lower charge-mass ratios. The electric force per unit mass on them is smaller. Accordingly, the target should be free of protons that would be preferentially accelerated [14], which puts strict requirements on experimental setups.

In this chapter, an efficient and simple method of heavy ion acceleration is proposed with the mechanism of electrostatic shock acceleration (ESA), where circularly polarized (CP) laser is used as the driver. As discussed in Chap. 1, the ponderomotive force of a CP laser does not contain an oscillating component, thus electrons in target are less heated than that for other polarizations. They are pushed forward and compressed by the light pressure, forming an intense self-consistent space-charge field that can accelerate the ions [15]. When the laser pulse is flat-topped, the perturbed system can propagate as an electrostatic shock. In the latter, the ions are mostly trapped and reflected forward at a nearly constant velocity, leading to the formation of a monoenergetic ion beam and a flat-top structure in the ion phase space [16]. In ESA, according to the scaling law of

L. Ji, *Ion Acceleration and Extreme Light Field Generation Based on Ultra-short and Ultra-intense Lasers*, Springer Theses,
DOI: 10.1007/978-3-642-54007-3_2,

Eq. (1.60), one can reduce the target density to obtain heavy ion bunches with higher energy. However, the target density must be above the relativistic critical value, so that the target remains opaque to the laser and the resulting space-charge field is sufficiently high to trap local ions [17].

Things become interesting when a compound target is used. It will be shown in the following simulations that, when the target consists of two ion species, both the heavy and light ions will be trapped and accelerated to the same velocity. A most important feature is that the common velocity is higher than when the target consists of only heavy ions, having the other parameters unchanged. This suggests a simple way to raise the energy of heavy ions: adding protons to the heavy ion target.

It should be noted that using compound or multilayer targets is not new in laser ion acceleration [3, 4, 12, 18–20]. But the heavy ions were more likely to be used as a background to increase the gradient of the space-charge field for proton acceleration. The corresponding laser pulses were usually linearly polarized. Furthermore, the resulting heavy ions from existing compound targets were not monoenergetic. To improve the quality of the accelerated heavy ions, in this chapter a sandwich target structure is employed. The latter consists of a thin compound ion layer between two light ion layers. In 3D, the target is micro-structured to yield ion beams of good quality. Simulation results show that at a laser intensity of $5 \times 10^{19}\,\mathrm{Wcm}^{-2}$ a carbon ion beam with energy divergence about 5 % at the longitudinal kinetic energy of $E_x \sim 58\,\mathrm{MeV}$, with a total carbon ion number of about 3.36×10^8 (a total charge of $3.23 \times 10^{-10}\,\mathrm{C}$) can be obtained.

2.2 CP Laser Interacting with Multispecies Target

First, a one-dimensional simulation is performed to observe the interaction between a relativistic CP laser pulse and a compound target, using the code VORPAL [21]. The light ion species is proton with charge and mass number of $\mathrm{Z}_1 = \mathrm{A}_1 = 1$ while the heavy one is carbon with $\mathrm{Z}_2 = 6$ and $\mathrm{A}_2 = 12$. A CP laser pulse with wavelength of $\lambda_0 = 1\,\mu\mathrm{m}$ irradiates the target from the left. The laser amplitude rises from zero to $a_L = 2$ in $6T_0$, where T_0 is the laser period, and then remains constant for $200\,T_0$. The simulation box is $101\,\lambda_0$ long in the x direction, and the cold target initially occupies the region between $x = 80.5\,\lambda_0$ and $x = 82.5\,\lambda_0$. The total electron density of the target is $n_e = 5n_c$, and the densities of the two ion species are given by $n_{e1} = n_{e2} = 0.5n_e$.

Figure 2.1a shows the phase space of the light and heavy ions, together with the electric field distribution. The presence of the collisionless electrostatic shock structure can be clearly seen. Almost all of both types of ions are trapped (reflected) by the electrostatic field, which is created through the laser-compressed electrons. In the region of the charge separating field, light ions seem to obtain prior acceleration and move ahead of heavy ions. However, all reflected ions move forward together when they leave the high-field region eventually. The light and heavy ions do not have different final velocities.

This interesting feature indicates that all ions and the accelerating field belong to the same self-consistent electrostatic shock system, whose speed is self-consistently determined by the laser plasma parameters, including the charge and mass of both ion species. Since the carbon mass is 12 times that of hydrogen, the energy per carbon ion is about 12 times that of protons, as can be clearly seen in Fig. 2.1b.

2.3 Analytical Modeling

The above simulation shows that the heavy and light ions are mostly trapped and accelerated to the same velocity. Assuming all the ions are accelerated, from momentum and energy conservation one obtains

$$\frac{I}{c} = -\eta\frac{I}{c} + \left(n_{e1}\frac{\mathrm{A}_1}{\mathrm{Z}_1} + n_{e2}\frac{\mathrm{A}_2}{\mathrm{Z}_2}\right) m_p v_i v_a, \tag{2.1}$$

$$I = \eta I + \left(n_{e1}\frac{\mathrm{A}_1}{\mathrm{Z}_1} + n_{e2}\frac{\mathrm{A}_2}{\mathrm{Z}_2}\right)\frac{m_p v_i^2}{2} v_a, \tag{2.2}$$

where I and η are the intensity and reflectivity of the laser pulse, m_p is proton mass, v_i and v_a are the velocities of the reflected ions and the shock, respectively, Z and A are ion charge and mass number, and $n_{e1} \equiv \mathrm{Z}_1 n_{i1}$, n_{i1} and $n_{e2} \equiv \mathrm{Z}_2 n_{i2}$, n_{i2} are the corresponding electron and ion densities of each species, respectively. The subscripts 1 and 2 denote the light and heavy ion species. Since the ions are reflected by the shock, the final velocity is about $v_i \approx 2v_a$, thus Eqs. (2.1) and (2.2) give

$$\frac{v_i}{c} \approx 2\sqrt{\frac{n_c}{n_{e1}\mathrm{A}_1/\mathrm{Z}_1 + n_{e2}\mathrm{A}_2/\mathrm{Z}_2}}\sqrt{\frac{m_e}{m_p}} a_L, \tag{2.3}$$

where n_c is the critical plasma density, m_e the electron mass, c the light speed, $a_L = eE_L/m_e\omega_L c$ the dimensionless laser electric field, e the elementary charge, and E_L and ω_L are the electric field amplitude and frequency of the laser. For a single-ion target, Eq. (2.3) reduces to Eq. (1.60).

To express it more clearly, Eq. (2.3) is rewritten in terms of the ratios $\alpha = n_{e1}/n_{e2}$ and $\beta = (\mathrm{Z}_1/\mathrm{A}_1)/(\mathrm{Z}_2/\mathrm{A}_2)$ as

$$v_i/c \approx 2\sqrt{1 + \frac{\alpha(\beta - 1)}{\alpha + \beta}}\sqrt{\frac{\mathrm{Z}_2}{\mathrm{A}_2}\frac{m_e}{m_p}\frac{n_c}{n_e}} a_L, \tag{2.4}$$

so that the effects of density ratio and different species on the acceleration efficiency become more transparent. For comparison, a series of simulations are carried out for various combinations of α and β values. The other conditions are

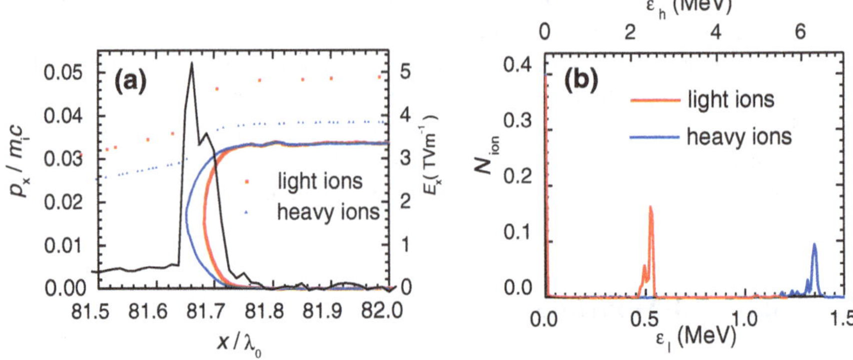

Fig. 2.1 **a** Phase space of the light (*red-dotted*) and heavy ions (*blue-dotted*) and the electric field distribution (*black solid*) of the collisionless electrostatic shock at $t = 150\,T_0$ for $a_L = 2$, $n_e = 5n_c$, and $d = 2\,\lambda_0$. **b** The energy spectrum of the light and heavy ions. ε_l and ε_h are the corresponding longitudinal kinetic energies, and $N_{\rm ion}$ is the normalized ion number. The compound target is a mixture of hydrogen (*light*) and carbon (*heavy*) with $n_{e1} = n_{e2} = 0.5n_e$

the same as in Fig. 2.1. The results are shown in Fig. 2.2, where an excellent agreement appears between the simulation and Eq. (2.4). Both simulations and analysis prove that the important feature exists for a very wide parameter range in shock acceleration. In Fig. 2.2a, one finds that the collective velocity in the compound target case is higher than that in the heavy ion only target case. That is to say, the energy of heavy ions can be easily increased by mixing with some light ions.

In order to improve heavy ion acceleration efficiently, Fig. 2.2a shows that its concentration must be much lower than that of proton. On the other hand, since the common velocity increases quickly with the density ratio at first and then slowly saturates to the maximum, the concentration of the heavy ions need not be reduced very much to achieve significant energy increment. This point is important to guarantee that the number of energetic heavy ions per bunch will not be too small. For example, for $n_{e1}/n_{e2} = 9$, or a carbon ion density of 10^{20} cm^{-3}, the velocity is $0.039\,c$, which is increased by 34.5 % with respect to the pure heavy ion target case, or 81 % by kinetic energy. It is close to the predicted maximum value of 41.5 % by Eq. (2.4). One should also note that the increasing rate of heavy ion energy does not depend on the other laser and target parameters as long as the shock acceleration mechanism dominates.

The results here stem from a self-consistent phase-mixing effect in the collisionless electrostatic shock. The local ions (electrons) are reflected, or accelerated, forward (backward) by the electrostatic space-charge field driven by the laser pulse. The light ions (with larger charge-mass ratio) tend to move faster than heavy ions. The faster (slower) ions will be pulled back (pushed forward), resulting in a redistribution of the space-charge-field. In such a manner, heavy ions experience stronger accelerating field while light ions less. Finally, the entire system moves

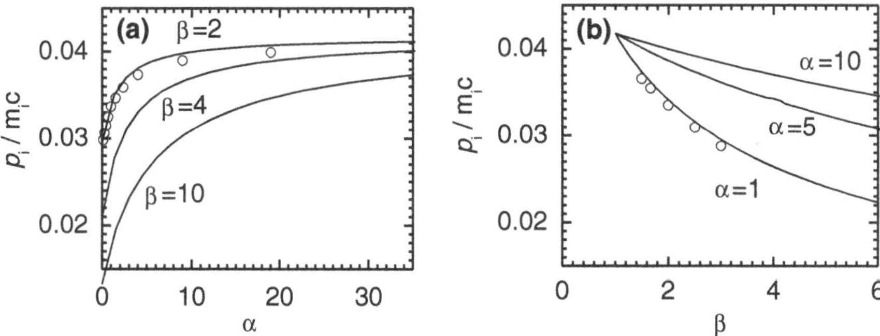

Fig. 2.2 Reflected heavy ion momentum *versus* **a** α and **b** β from simulations (*circles*) and Eq. (2.4) (*solid line*) at $t = 150\,T_0$. The other parameters are the same as in Fig. 2.1

together at a common speed. It seems that heavy ions are drawn by light ions indirectly, hence the light ions are slightly ahead of the heavy ions and a tiny valley exists in the electrostatic field distribution between the two reflected ion fronts in Fig. 2.1a. The circularly polarized laser pulse maintains the electrostatic field, whereas the response of the electron–ion plasma is another key of this interesting phenomenon. It should be mentioned that the higher the shock speed, the longer the time for picking up all heavy ions. PIC simulations also find that more complex multispecies targets (such as that with three species) exhibit similar phenomenon.

2.4 Generating Monoenergetic Heavy Ion Beam

In the above consideration, the laser pulse is flat-topped and the accelerated ions are monoenergetic. In most practical cases lasers are roughly Gaussian, so that according to Eq. (2.3), the accelerated ions may not be monoenergetic any more. The energy spread should be improved to fulfill the requirements of various applications.

2.4.1 "Sandwich" Target in One-Dimensional Simulation

To overcome the difficulty, a simple and effective method is presented. A sandwich target with a thin compound layer between two light ion layers is used, as seen in Fig. 2.3a. By this target structure, the beginning of the laser pulse would interact with the light ion layer in the front; then the following part with much smaller duration irradiates the thin compound layer, leaving the rest to the second light ion layer at the back. As the inside compound layer is very thin, its interacting

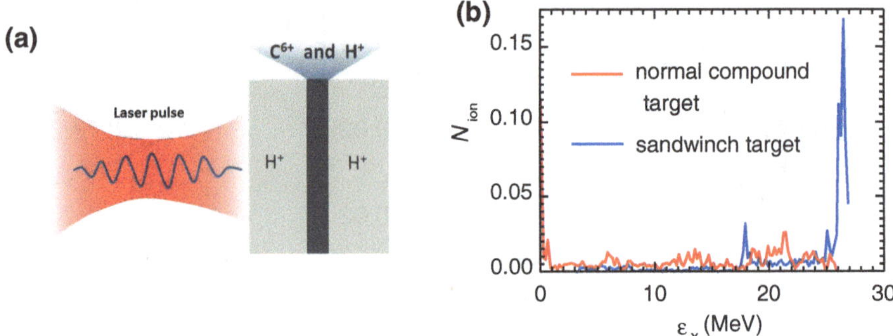

Fig. 2.3 Sandwich target scheme (**a**) and energy spectrum of normal compound target (*red solid*) and sandwich target (*blue solid*) (**b**) at $t = 240\,T_0$ for a Gaussian laser pulse with peak amplitude $a_L = 4$ and duration of $220\,T_0$. The inside central compound layer (hydrogen and carbon) is $0.2\,\lambda_0$ thick with $\alpha = 1$ and the outer two light ion layers (hydrogen) are both $0.9\,\lambda_0$ thick

time with partial laser pulse is short enough that the laser intensity keeps almost constant, which eventually leads to a more monoenergetic heavy ion bunch.

Here position and thickness of the thin layer have great influence on the beam quality. Taking Gaussian laser pulses for example, the intensity peaks in the center. It is better to place the compound layer in the right position of the target to make it interact with only the central part of the laser pulse, and thus ions would gain peak energy. Moreover, the energy spread may be manipulated by controlling the interacting time, which could be realized through the variation of the layer thickness. In a word, both the peak energy and the energy spread of the heavy ion beam can be regulated. A single thin compound foil cannot work like this because it might be destroyed by pre-pulse of the laser. It is also more convenient to produce this sandwich target than a single thin compound foil. The two light ion layers are attached to make the two ion species convenient to be separated in succeeding treatments.

For 1D simulations, the cold sandwich target electron density is $5n_c$ and the target is located from $x = 80.5\lambda_0$ to $x = 82.5\lambda_0$, while length of the simulation box is $101\lambda_0$. The central thin layer is a mixture of hydrogen and carbon ($\beta = 2$) with electron density ratio of $\alpha = 1$. Its thickness is $0.2\lambda_0$ and both the front and back hydrogen layers are $0.9\lambda_0$. The peak amplitude of the Gaussian laser pulse is $a_L = 4$ and its duration is $220\,T_0$. Comparison of ion energy spectrum with normal compound target is shown in Fig. 2.3b. All figures are taken at $t = 240\,T_0$ as the interaction is just over. An obvious improvement can be seen in the sandwich target way, by which carbon ions are almost around peak energy 26 MeV while the other is nearly averagely distributed. The parameters chosen here are to guarantee that the compound layer interacts only with the center part of Gaussian pulse in a short time, therefore heavy ions are well monoenergetic with peak energy. The peak energy is perfectly four times as in Fig. 2.1b according to the scaling law of Eq. (2.3), indicating that the acceleration scheme is well in operation.

2.4.2 *Microstructured Target in Three-Dimensional Simulation*

One may doubt the efficiency in practice because so far no 2D or 3D effects have been counted in. It has been confirmed by 2D PIC simulations in [15] that 2D effects do not have qualitatively influence on ion bunch formation, but it does affect energy and beam divergence of ion bunches. To weaken the negative effects, we reduced the transverse dimension of the inside thin compound layer. The target is microstructured with a compound microdot in it. Accordingly, a 3D PIC simulation by VORPAL is carried out. "1" and "2" denote hydrogen and carbon, respectively.

The simulation box is $25\lambda_0 \times 120\lambda_0 \times 120\lambda_0$. The microstructured target occupies a region of $15\lambda_0 - 17\lambda_0$ (2 μm) in x (longitudinal) and $-54\lambda_0 - 54\lambda_0$ (108 μm) in both $y$and z (transverse) with total electron density of $5n_c$. The inside microdot is mixed with hydrogen and carbon by $\alpha = 1$, occupying a region of $15.4\lambda_0 - 15.435\lambda_0$ (35 nm) longitudinally and $-2.4\lambda_0 - 2.4\lambda_0$ (4.8 μm) transversely. The rest of the target is made up of light ions. The Gaussian laser pulse with beam waist radius $w_0 = 17\lambda_0$, peak amplitude $a_L = 6$ and a duration of $40\,T_0$ propagates from the left. Figure 2.4a shows the proton (black) and carbon ion (red) distributions at $t = 30T_0$ where ions in the outer part of the target are eliminated to display the inside structure. The proton density shows a Gaussian-like front due to the Gaussian profile of the transverse laser intensity. All carbon ions in the compound layer have been reflected, leading to a compact beam with very small dimension. From phase space and longitudinal kinetic energy spectrum at different moment in Fig. 2.4b and c, the following process is shown clearly: the compound microdot is initially irradiated by laser front; carbon ions are being accelerated by the electrostatic field; carbon ions are all trapped and reflected, and finally propagating stably. This exactly describes the motion of the heavy ion from $t = 20\,T_0$ to $t = 32\,T_0$. In the end, aquasi-monoenergetic carbon ion bunch with peak energy of $E_x \sim 58$ MeV and energy spread of about 5 % is generated. The beam carries a total carbon ion number of about 3.36×10^8 (that is, a total charge of 3.23×10^{-10} C). Collimation of the obtained carbon ion bunch is pretty good as seen in Fig. 2.4d, with maximal divergence angle of $\Delta\theta_{max} \sim 6 \times 10^{-3}\,\pi$ rad.

The peak energy in Fig. 2.4c is about nine times as in Fig. 2.1b, verifying the scaling law again. One then concludes that 3D effect does not qualitatively impact acceleration mechanism either, and the proposed method turns out to be very effective. The transverse dimension of the microdot can be changed to control the energy divergence.

2.5 Summary and Discussion

In summary, a target with two ion species irradiated by a circularly polarized laser pulse is examined by PIC simulations. It is found that the two ion species are accelerated to the same velocity which is higher than in the case of the pure heavy

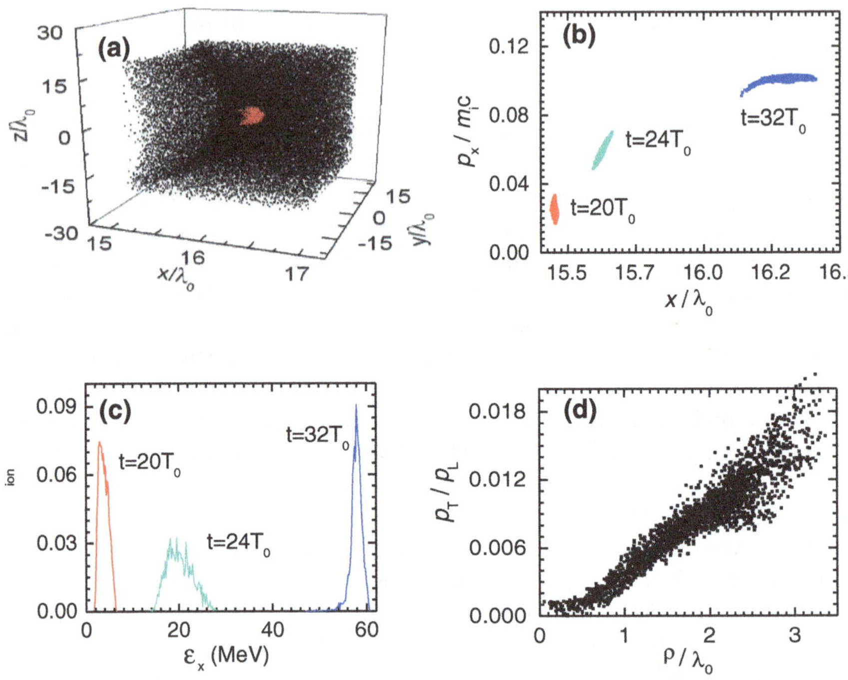

Fig. 2.4 Spatial distributions of hydrogen (*black dots*) and carbon ions (*red dots*) at $t = 30\,T_0$ (**a**); phase space (**b**) and energy spectrum (**c**) of heavy ions at different time and transverse distribution of p_L/p_T (**d**), where $p_L = \gamma m_i v_{ix}$ and $p_T = \gamma m_i (v_{iy}^2 + v_{iz}^2)^{1/2}$ are heavy ions' longitudinal and transverse momentum and $\rho = (y^2 + z^2)^{1/2}$ is the distance from the target center

ion target. A simple model based on momentum and energy conservations is proposed and it describes the found effect very well. When the laser pulse is not uniform in space and time, a sandwich microstructured target with a compound microdot in it is proposed. The 3D PIC simulation suggests that a quasi-monoenergetic heavy ion bunch can be generated. This method is so effective and practical that the energy of the heavy ions can easily be raised by nearly 100 % (or higher for heavier ions) under the same laser conditions. The qualities of the heavy ion beam, including the energy and space divergence, are significantly improved by using the microstructured target.

As an estimate, the LULI laser system (wavelength $\sim 1\,\mu$m, intensity $\sim 10^{19}\,$Wcm^{-2}, focal aperture $5 - 10\,\mu$m and duration about 300 fs) should be able to generate a carbon beam with a peak energy of 12 MeV and a total carbon ion number of $\sim 10^8$ while the energy divergence is still kept at 5 %. Moreover, the fraction of carbon ions in the target can be reduced such that a maximum accelerated energy of 18 MeV can be obtained.

References

1. M. Roth et al, Phys. Rev. Lett. **86**, 436 (2001)
2. R.A. Snavely et al., Phys. Rev. Lett. **85**, 2945 (2000)
3. H. Schwoerer et al., Nature **439**, 445–448 (2006)
4. B.M. Hegelich et al., Nature **439**, 441–444 (2006)
5. L.O. Silva et al., Phys. Rev. Lett. **92**, 015002 (2004)
6. J. Denavit, Phys. Rev. Lett. **69**, 3052 (1992)
7. Y. Sentoku et al., Phys. Plasmas **10**, 2009 (2003)
8. A. Zhidkov et al., Phys. Rev. Lett. **89**, 215002 (2002)
9. E.d'Humières et al., Phys. Plasmas **12**, 062704 (2005)
10. B. Shen et al., Phys. Rev. E **64**, 056406 (2001)
11. T. Esirkepov et al., Phys. Rev. Lett. **92**, 175003 (2004)
12. B. Shen et al., Phys. Rev. E **70**, 036403(R) (2004)
13. M. Hegelich et al., Phys. Rev. Lett. **89**, 085002 (2002)
14. B.M. Hegelich et al., Phys. Plasmas **12**, 056314 (2005)
15. A. Macchi et al., Phys. Rev. Lett. **94**, 165003 (2005)
16. X. Zhang et al., Phys. Lett. A **369**, 339–344 (2007)
17. X. Zhang et al., Phys. Plasmas **14**, 123108 (2007)
18. A.P.L Robinson et al., Phys. Rev. Lett. **96**, 035005 (2006)
19. M. Chen et al., Phys. Plasmas **14**, 113106 (2007)
20. T.Zh Esirkepov et al., Phys. Rev. Lett. **89**, 175003 (2002)
21. C. Nieter, J.R. Cary, J. Comp. Phys. **196**, 448 (2004)

Chapter 3
Ion Acceleration II: The Critical Target Thickness in Light Sail Acceleration

3.1 Introduction

Light pressure acceleration shows good prospect in generating monoenergetic ion beams with high efficiency, especially when it reaches the light sail (LSA) (or phase stable, PSA) regime. LSA has been well described in Chap. 1. It was shown that for certain laser intensity and plasma density, an appropriate target thickness is the most critical factor that defines whether the acceleration is in LSA/PSA regime. Normal estimations suggest that the critical foil thickness could be obtained by balancing the laser light pressure and the electrostatic force originating from charge separation. When the foil is thinner than the critical value, all electrons are expelled out of the target by light pressure and will disperse in a short period. No stable accelerating field will be formed.

In this chapter, we will re-examine the critical thickness issue. One-dimensional PIC simulations will demonstrate that the rising front of laser pulse has very important affection on the critical value. Normal estimation is true for laser pulses with sharp rising front. However, when the laser intensity rises much more slowly, the situation is quite different. A phase stable accelerating structure can still be well constructed even if the target thickness is below the critical one.

3.2 Estimation for Critical Thickness

We consider a circularly polarized laser pulse with normalized amplitude a_0 interacting with a foil of density n_0 and thickness D. Assuming the foil is opaque and the incident pulse is totally reflected, electrons pushed by the laser ponderomotice force pile up at the target back and form a dense skin layer, leading to a structure shown in Fig. 3.1. Based on this model, the critical foil thickness can be estimated via two approaches.

In the first approach, the balance condition of electrons located at the interface, i.e., at the position of $x = d$, is interpreted. They experience an electrostatic force

L. Ji, *Ion Acceleration and Extreme Light Field Generation Based on Ultra-short and Ultra-intense Lasers*, Springer Theses,
DOI: 10.1007/978-3-642-54007-3_3, © Springer-Verlag Berlin Heidelberg 2014

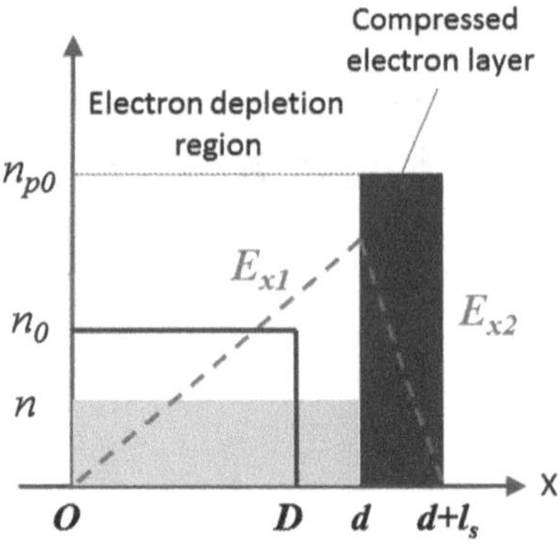

Fig. 3.1 Model of light sail (phase stable) acceleration [1]

from the maximum charge separation field $E_{//}$ at $x = d$. Meanwhile, since the transverse velocity of electrons v_e is close to the light speed c, the ponderomotive force may be simply represented by the Lorentz force, which is about $e|\mathbf{v}_e \times \mathbf{B}_0| \approx eE_0$. Here B_0 and E_0 are magnetic and electric fields of the incident laser, respectively. The critical thickness D is defined when all electrons are just expelled out to produce a maximally available charge separation field $E_{//} = 4\pi e n_0 D$ and the electrostatic force equals the Lorentz force $E_{//} = E_0$, which gives [1]

$$D = \frac{a_0}{n_0/n_c}\frac{\lambda_0}{2\pi}. \tag{3.1}$$

where λ_0 is the laser wavelength and n_c is the corresponding critical thickness.

The other way is to consider the balance of the whole electron layer located from $x = d$ to $x = d + l_s$, with l_s the skin depth. In this case, the light pressure acting on the layer is about $2I_0/c$, where I_0 is the peak laser intensity. The total electrostatic force is obtained by integrating eE_x over each electron, yielding $eE_{//}n_0D$. Compensating both forces produces another expression for critical thickness [2]

$$D = \frac{a_0}{n_0/n_c}\frac{\lambda_0}{\pi}. \tag{3.2}$$

An obvious difference from Eqs. (3.1) and (3.2) is that the latter is twice the former. In the first way, the Lorentz force takes the form of $e|\mathbf{v}_e \times \mathbf{B}_0| \approx eE_0$. This is true when the target is totally transparent or the foil is very thin so that laser could propagate through without any reflection. Electron at the interface then experiences the Lorentz force similar to a free electron in a laser field. However, in LSA the foil almost reflects the incident laser field by 100 %. The electric field at the reflecting surface is not simply the incident laser field E_0. The reflected field should be taken into account and thus more accurate theoretical analysis is required. Equation (3.2) is more reasonable and precise, which will be adopted in the following discussion.

3.3 Analysis on One-Dimensional Particle-in-Cell Simulations

To examine the validity of the former estimation, a series of 1D PIC simulations are carried out. The peak amplitude and wavelength of the simulated laser pulse is $a_0 = 20$ and $\lambda_0 = 1$ μm respectively, while the target density is $n_0 = 40n_c$. According to Eq. (3.2) the critical foil thickness should be $D = 0.16$ μm. In the first batch, protons are immobile. The foil thickness is set to be $l = 0.05$ μm, much smaller than the critical value. The laser pulse is with a plateau profile, i.e., laser amplitude linearly rises to maximum in a time t_{up} and then remains constant. Figure 3.2 displays the results with two different rising times. In Fig. 3.2a, the rising front is rather sharp with $t_{up} = 2\,T_0$ (T_0 is the laser period). One can see from the electron density distribution that all electrons are expelled out of the foil and then disperse, inducing a symmetric bipolar electrostatic field along the laser propagating direction. The laser pulse totally penetrates the foil and no stable structure of a positive accelerating field as in Fig. 3.1 appears. And of course LSA does not exist.

However, if the laser front rises more gently, say $t_{up} = 50\,T_0$, most electrons are well localized near the target back rather than being pushed out, as seen in Fig. 3.2b. A positive monopole electrostatic field is generated to balance the light pressure, which leads to stable acceleration once protons are allowed to move. Both results in Fig. 3.2 are taken when the flat-top part of the incident pulse is acting on the foil, guaranteeing the concordance of the comparing conditions.

One can conclude from the above analysis that the critical thickness requirement of Eq. (3.2) is not a necessity. The more essential criterion lies in Ref. [3]. For laser pulse with a gentle rising, hydrodynamics is valid and there is always a stable solution for present parameters, indicating that a stable positive monopole electrostatic field would be generated and the electron layer can be well confined. While for a fast varying laser field such as a several-cycle laser pulse, as discussed in Ref. [4], the gradient of pondermotive force in the electron layer is so large that not only are electrons displaced but kinetic effect becomes important. Thus, one will not gain a stable hydromechanical solution. It cannot afford a stable electrostatic field and electrons will be all pushed out and disperse, as mentioned in Fig. 3.1a, leading to no phase stable acceleration.

Therefore the new criterion is, for a given parameters, as long as a stable solution exists, LSA can be achieved even the target is much thinner.

In Fig. 3.2b, it seems laser field mostly penetrates through and hence the acceleration efficiency is not high. Nevertheless target transparency will decrease a

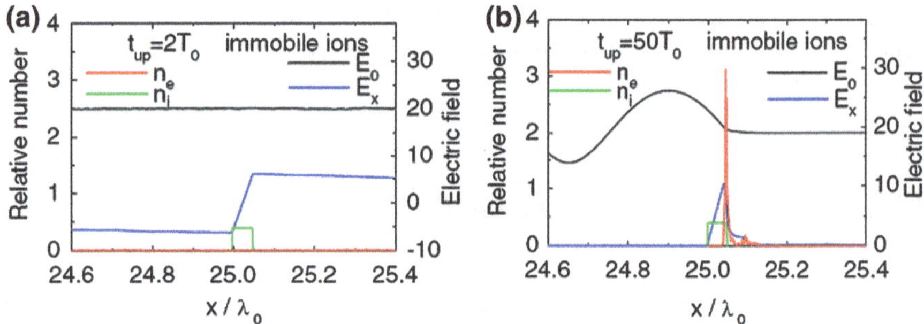

Fig. 3.2 Distributions of laser field E_0, longitudinal electric field E_x, electron density n_e and ion density n_i with laser rising front of $t_{\text{up}} = 2T_0$ (**a**) and $t_{\text{up}} = 50T_0$ (**b**). Here protons are immobile

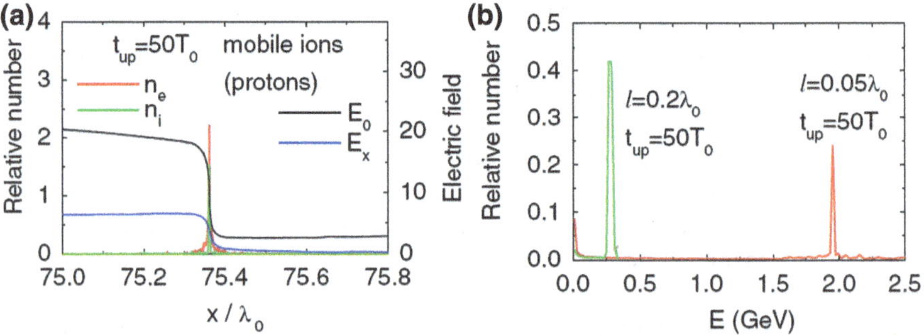

Fig. 3.3 Distributions of laser field E_0, longitudinal electric field E_x, electron density n_e and ion density n_i with laser rising front of $t_{\text{up}} = 50T_0$ and mobile protons (**a**). Proton energy spectrums of different target thickness of $l = 0.05$ μm and $l = 0.2$ μm (**b**). Electric field is normalized to dimensionless amplitude a

lot with mobile protons. The transmitted laser field in Fig. 3.3a is nearly one-tenth of the one in Fig. 3.2b. To see this through, let us consider the target being compressed from l to d. The laser field decays in the form of e^{-x/λ_s} in overdense plasma. Here $\lambda_s = c/\sqrt{4\pi n_e e^2/m_e}$ is the temporal plasma skin depth and n_e is the instantaneous foil density. According to charge conservation law, one has $n_e d = n_0 l$. So

$$\frac{\lambda_s}{d} = (\frac{\lambda_{s0}}{l^{1/2}})d^{-1/2} \sim [\ln(E_0/E_t)]^{-1} \tag{3.3}$$

can be taken as the laser transparency, representing the ratio between the transmitted laser field E_t and the incident laser field E_0. As electron and proton layers move forward in a speed of v, the light pressure in its moving frame shall include the Doppler effect $P_L = \frac{E_0^2}{2\pi}\frac{c-v}{c+v}$ [2]. As v increases with time, the light pressure P_L decreases and the compression of the electron layer is weaker, making its thickness

d increase. Accordingly, the value of λ_s/d decreases, meaning that the target becomes more and more opaque and laser penetration is much less than in the case of immobile protons. As the foil is continuously accelerated, the transparency will be lower and lower and high-energy efficiency is achieved.

The relaxation of the limit on foil thickness from Eq. (3.2) is of practical significance. As introduced in the first chapter, the LSA becomes not efficient when the protons move at relativistic velocities and the energy increases gradually with accelerating distance and time. The conclusion in this chapter suggests that acceleration efficiency can be enormously increased with target thickness much lower than the so-called critical value. The energy spectrum comparison of $l = 0.05$ μm and $l = 0.2$ μm is shown in Fig. 3.3b. The peak energy of $l = 0.05$ μm is about eight times the other after depleting the same laser energy.

Using ultra-thin foils can to some degree solve the inefficient LSA problem in the relativistic regime.

3.4 Summary and Discussion

In this chapter, the critical thickness in LSA is clarified via 1D PIC simulations. The results show that the laser rising front has crucial effect on the issue. When the laser intensity rises gently, even the foil thickness is well below the so-called critical value, stable accelerating structures can still be maintained as long as stationary solution exists. Thinner foils offer higher peak energy with the same laser conditions.

In reality, multidimensional effects should be interpreted. Various instabilities may also play important roles, which is quite a challenge and requires more concrete studies.

References

1. X.Q. Yan et al., Phys. Rev. Lett. **100**, 135003 (2008)
2. T.Zh. Esirkepov et al., Phys. Rev. Lett. **92,** 175003 (2004)
3. B. Shen et al., Phys. Rev. E **64**, 056406 (2001)
4. B.F. Shen et al., Phys. Plasmas **8**, 1003 (2001)

Chapter 4
Extreme Light Field Generation I: Quasi-Single-Cycle Relativistic Laser Pulse

4.1 Introduction

Recent significant improvements in laser light contrast by means of the double plasma mirror [1] and other techniques [2–5] allow an intense laser pulse to interact with a solid-density foil before the latter is damaged by the prepulse of the laser. Very hard diamond-like foils of ultrasmall, say 4.5 nm, thickness comparing with the skin depth of the light are now available. All these offer a good condition for studying relativistic laser–foil interaction.

Target has always being the main concern in the regime of ultra-intense laser interacting with overdense plasma. Most researches have been dedicated to particle acceleration, where several important mechanisms were proposed as mentioned in Chaps. 1 and 2. On the other hand, if one takes the laser field as the studying subject, even more interesting phenomena and principles would show up, which may also induce many crucial applications. The "relativistic oscillating mirror" model introduced at the end of Chap. 1 turns to be a typical example. Another application is to use an ultrathin foil as a relativistic mirror for generating high-intensity ultrabright X- and gamma-ray radiation by relativistic Doppler shifting the light [6]. Laser energy can also be trapped and accumulated to very high levels between two closely placed ultrathin foils when two oppositely directed ultraintense laser pulses impinge on them [7].

In this chapter, a new method of producing a nearly single-cycle ultraintense light pulse is proposed via laser–foil interaction. Single-cycle lasers are suitable for generating single attosecond pulses [8] as well as for electron acceleration in the small bubble regime [9, 10]. Several methods to obtain single-cycle laser pulses have been proposed, most of which are optical [11]. However, owing to the relatively low damage threshold of the optical components and other problems, the intensity of the optically produced single- and few-cycle laser pulses is very limited.

The following sections will show by analytical modeling and PIC simulations that, when a laser light interacts with an adequately thin overdense plasma, both are self-consistently (nonlinearly) modulated. For suitably chosen laser and foil

L. Ji, *Ion Acceleration and Extreme Light Field Generation Based on Ultra-short and Ultra-intense Lasers*, Springer Theses,
DOI: 10.1007/978-3-642-54007-3_4,

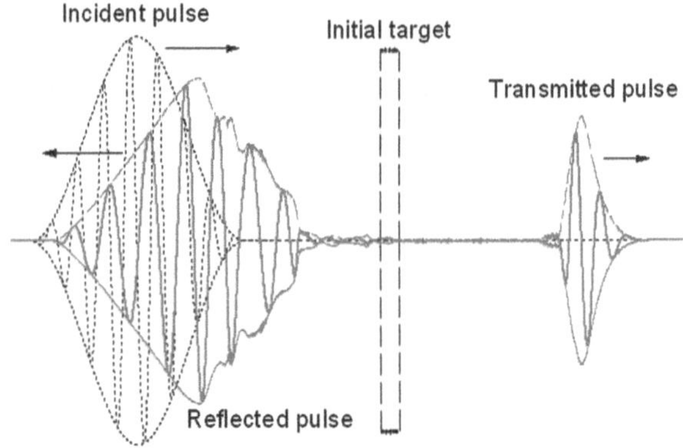

Fig. 4.1 Scheme for generating a near-single-cycle laser pulse. The incident pulse interacts with a thin foil target, producing an ultrashort transmitted pulse and a reflected pulse

parameters, the foil is transparent only to the highest intensity part of the laser pulse. As a result, the transmitted light pulse has a much smaller duration than the incident pulse, as shown in Fig. 4.1. The pulse duration depends primarily on the laser light intensity gradient and foil conditions. Since the process involves only laser–plasma interaction, like the plasma grating [12–14], there should be no intensity limitation due to material damage.

4.2 Quasi-Single-Cycle Pulse from Laser (CP)–Foil Interaction

We shall first perform a one-dimensional (1D) particle-in-cell (PIC) simulation with LPIC++ [15]. The normalized amplitude $\boldsymbol{a} = e\boldsymbol{E}_0/m_e\omega_0 c$, where e and m_e are the electronic charge and mass, respectively, $\boldsymbol{E}_0$ is the laser electric field, ω_0 is the laser frequency, and c is the speed of light, of the circularly polarized (CP) incident laser pulse, propagating in the x direction, is given by $\boldsymbol{a} = a_m \sin^2(\pi t/2\tau)$ $[\sin(\omega_0 t)\hat{e}_y + \cos(\omega_0 t)\hat{e}_z]$, where $a_m = 20$ is the peak amplitude and $\tau = 4$ is the pulse duration normalized by the laser period T_0. The laser light is of wavelength $\lambda_0 = 1$ μm. The simulation box is $50\lambda_0$ in the x direction, and the foil is initially located between $x = 20\lambda_0$ and $20.7\lambda_0$. The normalized (by the critical density $n_c = m_e\omega_0^2/4\pi e^2$) foil density is $N_0 = 8$. The cold foil plasma is taken to be fully ionized and the ion-to-electron mass ratio is $m_i/m_e = 1,836$. The simulation mesh size is $\lambda_0/200$.

Figure 4.2a shows the trajectories of 71 electrons and 71 protons that are initially uniformly distributed in the foil. The laser field behind the foil is also recorded simultaneously. The incident pulse arrives at the foil at $t = 20T_0$ and its rising part pushes the foil electrons inward. An electron layer is formed and compressed by the laser ponderomotive force. Laser light transmission through the foil occurs at about $t = 23T_0$. As the electron layer is further compressed, the transmitted laser field increases rapidly and reaches maximum at about $t = 25T_0$,

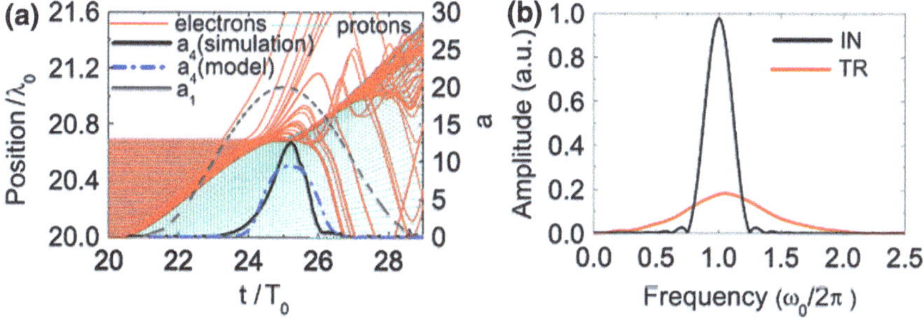

Fig. 4.2 PIC simulation results for $a_m = 20$, $\tau = 4$, and $d = 0.7$ (normalized by laser wavelength). **a** Electron and proton trajectories, the laser field a_4 at the foil backside from simulation and analytical model and incident laser field a_1. **b** Spectrums of the incident (*IN*) and the ultrashort transmitted (*TR*) light pulses

while the compression also peaks. When the trailing part of the incident pulse enters the foil plasma, the foil protons start to move forward with the electron layer because of the space charge field. The transmitted field also drops sharply from its peak value, resulting in an ultrashort, nearly single-cycled transmitted light pulse with duration $1.1T_0$ and peak amplitude $a = 12.5$. That is, at the small cost of amplitude reduction (from 20 to 12.5), the duration of the incident pulse is narrowed to about one-fourth of the incident pulse. One can also see that as the peak of the incident pulse enters the plasma, ion motion becomes significant. This can be attributed to the fact that at this stage the charge-separation field has become sufficiently strong. The electrons and ions are then driven forward by the laser as a double layer. The spectrums of incident and transmitted laser pulses are shown in Fig. 4.2b. One can see that the light spectrum is broadened by about three times because of the pulse shorting. The central frequency is barely changed, showing that there is almost no frequency shift and the number of light wave cycles is indeed reduced by the pulse shortening.

4.3 Nonlinear Modulation of Foil Transparency

4.3.1 Stationary Solution

We now consider in more detail the physics of the pulse compression process. Figure 4.3 shows a model of the laser–foil interaction [16]. In the model the electron dynamics in the laser–foil interaction is treated as quasi-static in the sense that as the laser field varies with time, for each value of a, the corresponding stationary solution is obtained. This approximation is valid for overdense ($N_0 = 8$) plasmas, since the plasma response time $\omega_{pe}^{-1} \sim T_0/N_0^{1/2}$ is much smaller than pulse duration $\tau(= 4T_0)$. In this model the ions remain stationary.

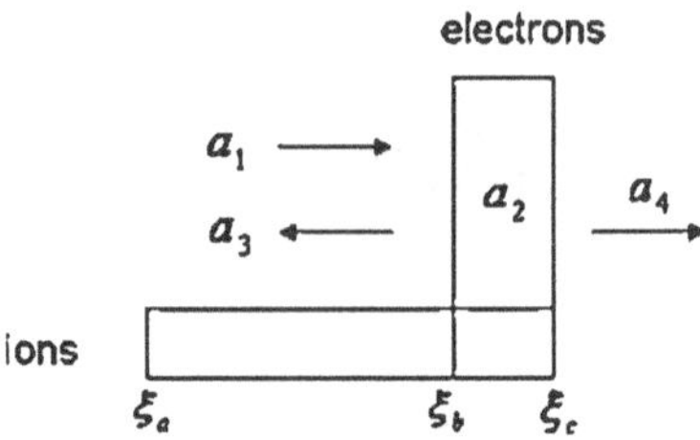

Fig. 4.3 The laser–foil interaction model. The incident, reflected, transmitting, and transmitted laser fields are denoted by a_1, a_2, a_3, and a_4, respectively. The initial left and right surfaces of the foil are at ξ_a and ξ_c. The interface of the laser front and the electron layer is at ξ_b

The normalized vector potential of laser field at $\xi = \omega_0 x/c$ can be written as $a = a_0(\xi)\exp(i\omega_0 t + i\theta(\xi))$, where $\theta(\xi)$ is the wave phase. Two constants of motion of an electron moving in the laser field are [17]

$$M = (\gamma^2 - 1)\partial_\xi \theta, \tag{4.1}$$

$$W = \left[(\partial_\xi \gamma)^2 + M^2\right]/2(\gamma^2 - 1) + \gamma^2/2 - N_0, \tag{4.2}$$

where $\gamma = (1 + a^2)^{1/2}$ is the relativistic factor. Requirement of stationary solutions gives [15]

$$\partial_\xi \gamma = \partial_\xi \psi \equiv -E_x, \tag{4.3}$$

where $\psi = e\phi/m_e c^2$ is the normalized scalar potential and $E_x = -\partial_x \phi$ is the electrostatic charge-separation field. Invoking continuity of the transverse electric and magnetic fields at the interfaces ξ_b and ξ_c, we have

$$a_4^2 = a_c^2 = M,\ \partial_\xi a_4 = 0,\ \text{and}\ W = |M| + 1/2 - N_0(1 + |M|)^{1/2}, \tag{4.4}$$

$$(\partial_\xi a_b)^2 + M^2/a_b^2 + a_b^2 = 4a_1^2 - 2M, \tag{4.5}$$

$$\left[(\partial_\xi \gamma_b)^2 + M\right]/2(\gamma_b^2 - 1) + \gamma_b^2/2 - N_0\gamma_b = W. \tag{4.6}$$

The initial foil thickness is $\xi_{ac} = \xi_{ab} - \xi_{bc}$, while $\xi_{ab} = |E_{xb}|/N_i$. According to Eq. (4.3), one obtain $\xi_{bc} = \int_{\gamma_b}^{\gamma_c} (-1/E_x)d\gamma$. Here the subscripts represent the values of each variable at corresponding positions. The numerical solution is obtained through the following process [16]: (a) assign a series of values for the invariant M; (b) Substitute each M value into Eqs. (4.4)–(4.6) and calculate the corresponding foil thicknesses of ξ_{ac}; (c) select the appropriate M value that produces the given foil thickness; (d) substitute the right M value to all equations and obtain each variable.

4.3.2 Modulation Mechanism and Parametric Study

In order to understand the modulation mechanism of foil transparency, a relatively thick plasma target of $d = 4$ is used with immobile ions, while the other parameters are the same as that in Fig. 4.2. The simulated electron trajectories are shown in Fig. 4.4a. Through imbalance of the ponderomotive and the electrostatic charge-separation forces, the electrons are steadily pushed inward as well as reflected, corresponding to the rising and falling regions of the incident pulse. The relative position (ξ_{ba}) of the electron layer versus time calculated from our analytical model is also given. It describes the motion of the plasma boundary very well.

An important parameter in the scheme is the maximum displacement X_m of the electron-layer surface. It is associated with the peak laser amplitude, calculated to be $X_m \sim 0.77\lambda_0$, as seen in Fig. 4.4a. This value also agrees with that from the simulation. When the foil is too thick, say $d = 4$, no transmission occurs. If the foil thickness is reduced to around X_m, though still larger than the skin depth, significant laser-light transmission takes place due to compression of the electron layer. The laser field decays as e^{-x/λ_s} in an overdense plasma, where $\lambda_s = c/\omega_{pe} = (\gamma_b/N_0)^{1/2}(l/d)^{1/2}\lambda_0/2\pi$ is the skin depth. When the electron layer in a foil of thickness d is compressed to thickness l, the parameter $\alpha = \lambda_s/l$: $1/\ln(E_0/E_t)$ can be used to represent the foil transparency. Accordingly, one has

$$\alpha = \frac{\lambda_0}{2\pi}\left(\frac{\gamma_b}{N_0}\frac{1}{d}\right)^{1/2}\frac{1}{l^{1/2}}, \tag{4.7}$$

so that if the skin depth is larger than the layer thickness, i.e. $\alpha > 1$, laser transmission would be significant. No transmission occurs if $\alpha << 1$. To see how the transparency is related to the laser amplitude, we have calculated α versus time for three different initial foil thicknesses around X_m. It can be clearly seen in Fig. 4.4b that for all the three thicknesses, α first increases with rising laser amplitude, reaches a maximum at the peak incident-pulse amplitude, and then decreases with the falling part of the laser pulse. That is, the foil transparency is modulated by laser amplitude: the less intense part of the pulse is reflected and the more intense part can easily pass through. As a result, the transmitted light pulse is much shorter than the incident pulse and an ultrashort laser pulse is produced. Figure 4.4b also shows the effect of the foil thickness. In the case $d = 0.8$, since the peak value of α is rather small (~ 1), the transmitted pulse is of low intensity. A thinner foil such as $d = 0.6$ yields larger peak transparency, but the duration of the resulting pulse is also longer. The case $d = 0.7$ presented in Fig. 4.2 offers both short duration and large amplitude. Figure 4.5 shows the relationship between parameters of the transmitted pulse and target thickness. One can choose appropriate target thickness for specific application requirements. We should notice that with increasing target thickness, the peak amplitude of the transmitted pulse decreases all along while its duration decreases rapidly to about one cycle at $d = 0.7$ and becomes saturated. It

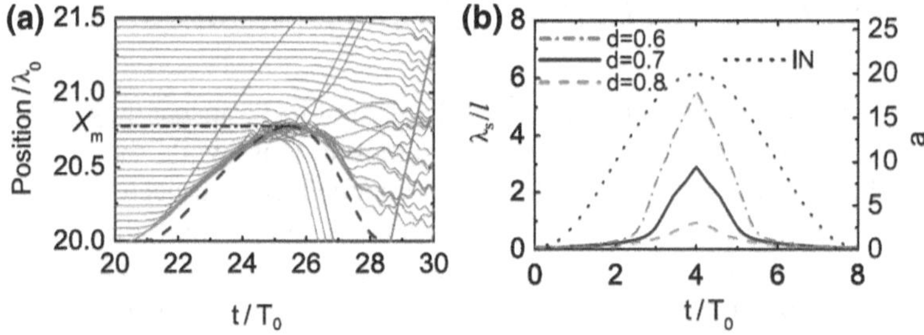

Fig. 4.4 **a** Electron trajectories (*solid*) from 1D PIC simulations for immobile ions and the calculated interface position $\xi_{ba} = \xi_b - \xi_a$ (*dashed*) versus time for $a_m = 20$, $\tau = 4$, $N_0 = 8$, and $d = 4$. The maximum displacement of the interface is marked by X_m. **b** Calculated value of λ_s/l versus time for $d = 0.6$ (*dash-dot*), 0.7 (*solid*), and 0.8 (*dashed*). The pulse shape of the incident laser (*dotted*) is also shown

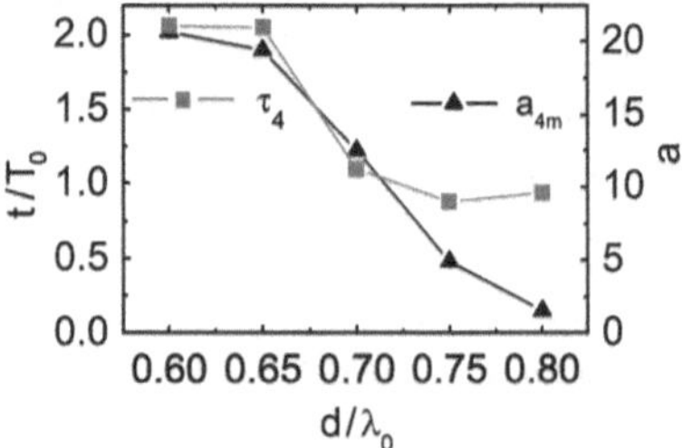

Fig. 4.5 The peak amplitude a_{4m} (*triangle*) and duration τ_4 (*square*) of the generated pulse versus target thickness with $a_m = 20$, $\tau = 4$, and $N_0 = 8$ from simulations

means that $d = 0.7$ corresponds to a pulse with largest peak amplitude in nearly single-cycle region, which is consistent with the analysis in Fig. 4.4b.

The temporal profiles of the generated laser pulses from the simulation and the analytical model for $d = 0.7$ are also compared in Fig. 4.2a. We see that the agreement is quite good. A difference is that the ultrashort pulse from the simulation is not symmetrical, with the tail part steeper than the rising front. This can be attributed to ion motion. As mentioned, when most of the incident pulse has entered the target, the ions and electrons are driven forward as a double layer. If the latter has a velocity v, the light, or ponderomotive, pressure P_L on it can be written as

$$P_L = \frac{E_0^2}{2\pi} \frac{c - v}{c + v}, \tag{4.8}$$

where the Doppler effect has been included. According to Eq. (4.8), the foil velocity v reduces the light pressure by a factor $(c - v)/(c + v)$ and thus weaken the compression of the layer. The layer thickness l would then be larger than that in the immobile-ion case of the analytical model. From Eq. (4.7) we have $\lambda_s/l \sim l^{-1/2}$, so that the foil transparency decreases with larger l. In this manner, the sudden increase in foil velocity leads to the shortened pulse tail. Even if the incident pulse has a longer tail it would be cut-off by this effect. In present simulation, the layer

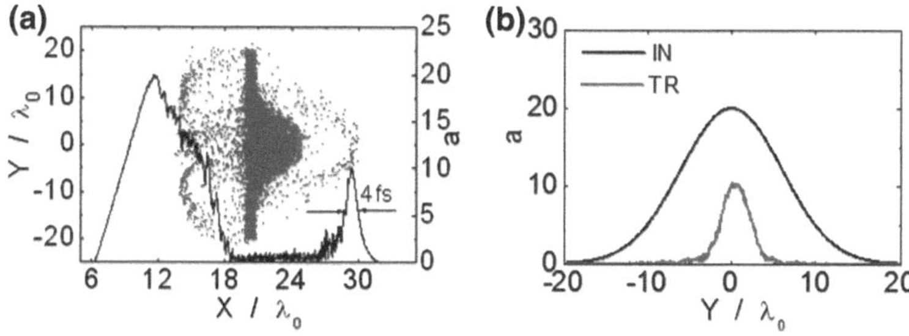

Fig. 4.6 Results from the 2D PIC simulation. The laser parameters are $a_m = 20$, $\tau = 4$, and the transverse FWHM is $12\lambda_0$. The foil density is $N_0 = 8$ and the thickness is $d = 0.75$. **a** Electron distribution (*dots*) and axial laser profiles at $Y = 0$ (*solid*). **b** Transverse profiles of the incident (*IN*) and transmitted (*TR*) laser pulses

velocity is about $v/c \approx 0.2$ as the peak laser field interacts with the foil, which decreases the light pressure by nearly 50 % according to Eq. (4.8).

Some electrons at the foil backside are driven away by the electric field of the transmitted pulse directly, which has not been included in our analytical model, therefore the transparency of the target (hence the peak amplitude of transmitted pulse) in simulation is somewhat higher, as can be seen in Fig. 4.2a.

It is somehow difficult to tell analytically how large the amplitude gradient shall be to gain a quasi-single-cycle pulse, because the duration is also related to target thickness. However, for each a_m and τ, one can always follow the above procedure to gain the relationship between transmitted and incident pulses with different target thickness, as seen in Fig. 4.5, and consequently choose the appropriate parameters.

4.4 Two-Dimensional Simulation

2D PIC simulations of the proposed scheme are carried out to verify the mechanism in a higher dimension. The simulation box is $50\lambda_0 \times 50\lambda_0$ in the X and Y directions. The $N_0 = 8$ cold-plasma foil occupies a region from $20\lambda_0$ to $20.75\lambda_0$ in the X direction and $-20\lambda_0$ to $20\lambda_0$ in the Y direction. The mesh size is $(\lambda_0/60) \times (\lambda_0/60)$. A CP laser pulse, with $a_m = 20$, $\tau = 4$, and transverse FWHM $12\lambda_0$, illuminates the foil from the left. The foil thickness is slightly increased (comparing to the 1D case) to $d = 0.75$ in order to compensate the foil deformation by hole boring. Figure 4.6a shows the electron distribution and axial laser profiles at $Y = 0$. The transverse profile of the incident and transmitted pulses at the position of the peak intensity in the X direction are shown in Fig. 4.6b. A distinctive ultrashort laser pulse is well generated. The pulse duration is $\sim$4 fs and the peak intensity is $\sim 3 \times 10^{20}$ W/cm^2, comparable to the 1D results (3.7 fs, 4.3×10^{20} W/cm^2). The transverse FWHM of the transmitted pulse, which is determined mainly by the focal spot size of incident pulse, is $\sim 4\lambda_0$. That is, the transmitted pulse is localized in a $1.2\lambda_0 \times 4\lambda_0$ region. Thus, the present scheme offers a possibility for producing a high-intensity λ_0^3 laser pulse [18]. In Fig. 4.6a, there are still some breaking away electrons co-moving with the transmitted pulse. Their energy is around tens of MeV, thus can be separated from the generated pulse through a one Tesla magnetic field.

4.5 Summary

In summary, using PIC simulations and analytical modeling, we have shown that a nearly single-cycle relativistic laser pulse can be obtained when a CP laser pulse interacts with a dense foil. The scheme is based on the fact that the transparency of the foil plasma is modulated by the laser. Only a small segment associated with the most intense part of the latter is transmitted, the rest are reflected and absorbed. The results from the simulation and the quasi-static model agree well, and the dependence on the transparency parameter that controls the intensity and width of the transmitted pulse is investigated. The main conclusions are also verified by the 2D PIC simulation. The pulse duration mainly depends on the gradient of the incident-pulse profile. With a longer laser pulse, one may use multiple foils to reduce the pulse duration step-by-step and obtain quasi-single-cycle transmitted pulses finally. This scheme deals with submicron foil, hence requires high contrast laser pulse, and recent progress in ultrahigh contrast laser pulse techniques such as Ref. [1–5] offers feasibility of the proposed scheme.

References

1. T. Wittmann et al., Rev. Sci. Instrum. **77**, 083109 (2006)
2. A. Jullien et al., Opt. Lett. **30**, 920 (2005)
3. V. Chvykov et al., Opt. Lett. **31**, 1456 (2006)
4. D. Homoelle et al., Opt. Lett. **27**, 1646 (2002)
5. R. Shah et al., Opt. Lett. **34**, 2273 (2009)
6. T.Zh. Esirkepov et al., arXiv: 0812.0401 (2008)
7. B. Shen et al., Phys. Rev. Lett. **89**, 275004 (2002)
8. T. Brabec, F. Krausz, Rev. Mod. Phys. **72**, 545 (2000)
9. A. Pukhov et al., Appl. Phys. B **74**, 355 (2002)
10. K. Schmid et al., Phys. Rev. Lett. **102**, 124801 (2009)
11. F. Tavella et al., Opt. Lett. **32**, 2227 (2007)
12. H.C. Wu et al., Phys. Plasmas **12**, 113103 (2005)
13. H.C. Wu et al., Appl. Phys. Lett. **87**, 201502 (2005)
14. H.C. Wu et al., Laser Part. Beams **23**, 417 (2005)
15. R. Lichters, R.E.W. Pfund, J. Meyer-ter-Vehn, LPIC ++, Rep. No. MPQ225, Max-Planck-Institut fuer Quantenoptik, Garching (1997)
16. B. Shen et al., Phys. Plasmas **8**, 1003 (2001)
17. C.S. Liu, Phys. Rev. Lett. **36**, 966 (1976)
18. G. Mourou et al., Plasma Phys. Rep. **28**, 12 (2002)

Chapter 5
Extreme Light Field Generation II: Short-Wavelength Single-Cycle Ultra-Intense Laser Pulse

5.1 Introduction

In this chapter, a plasma approach of generating intense chirped pulse is presented via particle-in-cell (PIC) simulations. When a circularly polarized (CP) laser pulse, the driving pulse (DR pulse), interacts with an overdense foil, the whole electron and proton layer is accelerated by light pressure and moves at an ultra-high velocity. With another counter-propagating CP pulse named as the scattered pulse (SC pulse) impinging on this high-speed double layer, the SC pulse would be reflected and hence its frequency is strongly Doppler shifted. The amount of the frequency shift is determined by the layer velocity which varies with time, so the SC pulse is highly chirped. Its chirped component can be manipulated by controlling the layer velocity versus time, which is realized through shaping the DR pulse. Once group velocity dispersion (GVD) is roughly compensated, such as inserting materials in the laser beam, the chirped SC laser is compressed to a nearly single-cycled relativistic pulse with a much shorter wavelength.

Compared to the traditional optical approaches, this method has several advantages. First, the intensity of the chirped pulse is extremely high because only laser-plasma interaction is involved. Second, by controlling the chirped component, the spectrum is much broader than traditional optical approaches. After compression, the pulse intensity increases even more. Lastly the wavelength of the reflected pulse is much shortened from the Doppler shifting.

5.2 Intense Chirped Pulse Generated by Double-Sided Laser (CP)–Foil Interaction

As usual, to give a simple map about the scheme, a one-dimensional PIC simulation is performed by the code VORPAL [1]. A CP pulse, with wavelength $\lambda_0 = 1\ \mu\text{m}$ and peak amplitude $a_0 = 50$, irradiates on a thin foil as the driving pulse. Here $\boldsymbol{a} = e\boldsymbol{E}_L/m_e\omega_0 c$ is the normalized dimensionless amplitude, where

L. Ji, *Ion Acceleration and Extreme Light Field Generation Based on Ultra-short and Ultra-intense Lasers*, Springer Theses,
DOI: 10.1007/978-3-642-54007-3_5,

e and m_e are the electronic charge and mass, respectively, $\boldsymbol{E}_L$ is the laser electric field, ω_0 is the laser frequency, and c is the speed of light in vacuum. The driving pulse has a plateau shape, with a laser field rising from zero to $a_0 = 50$ in $5T_0$ (T_0 is the laser period) and then remaining constant. Meanwhile a counter-propagating CP pulse, with peak amplitude $a_{sc} = 5$ and FWHM (full-width half-maximum) duration $\tau_{sc} = 30T_0$, impinges on the foil from the other side as the scattered pulse. The foil density is $n_0 = 100n_c$ and its thickness is $d = 0.2\lambda_0$ where $n_c = m_e\omega_0^2/4\pi e^2$ is the critical density. The simulation box is $100\lambda_0$ long in the x direction (propagating direction), and the foil is initially located between $x = 48\lambda_0$ and $48.2\lambda_0$. The simulation mesh size is $\lambda_0/250$.

Figure 5.1 shows the simulation results. Figure 5.1a displays the laser field distribution after the whole interacting process. One could see that the whole foil is accelerated by the DR pulse in the LPA (PSA) scheme. The SC pulse, propagating in the opposite direction, interacts with the foil subsequently and then is totally reflected. Its frequency is Doppler shifted, which is determined by the foil velocity. The temporal frequency of the reflected laser field increases with the double-layer velocity. As a result, a positively chirped pulse is generated, as seen clearly in Fig. 5.1b. The reflected field of the DR pulse is also chirped, of course, negatively. After being reflected from the fast moving layer, the duration of SC pulse is compressed by a factor of 5. The spectra before and after reflected are compared in Fig. 5.1c. The central frequency is shifted by a factor about 5, consistent with pulse compression. An important feature is that FWHM of the spectrum is enormously broadened, reaching $2.6\omega_0$. This effect is mainly caused by the pulse being chirped. Figure 5.1c also shows phase information. It is seen that difference between the full dispersion phase and the second-order dispersion phase can be neglected, which means that second-order dispersion is dominant and the chirp is almost linear. After full dispersion compensation, a nearly single-cycled laser pulse with duration of 0.24T_0 (0.8 fs) is obtained, as seen in Fig. 5.1d.

Meanwhile, the laser wavelength is much shortened to about one-fifth of the original pulse (0.2 μm). Its cycle number is also enormously decreased. The peak dimensionless amplitude of the scattered pulse is increased from $a_0 = 5$ (normalized by the initial fundamental frequency) to $a_0 \approx 55$ (normalized by the central frequency of the compressed pulse itself) after compression, reaching strong relativistic region. Note that the dimensionless amplitude a does not change while only Doppler shift is considered. In other words, we have gained a relativistic single-cycled pulse with short wavelength which has not been realized with any method but is believed to be very useful in many applications. It should be mentioned that if only second-order dispersion is compensated, the pulse duration also reaches 0.36T_0, very close to the case of full dispersion compression since the generated pulse is quasi-linearly chirped.

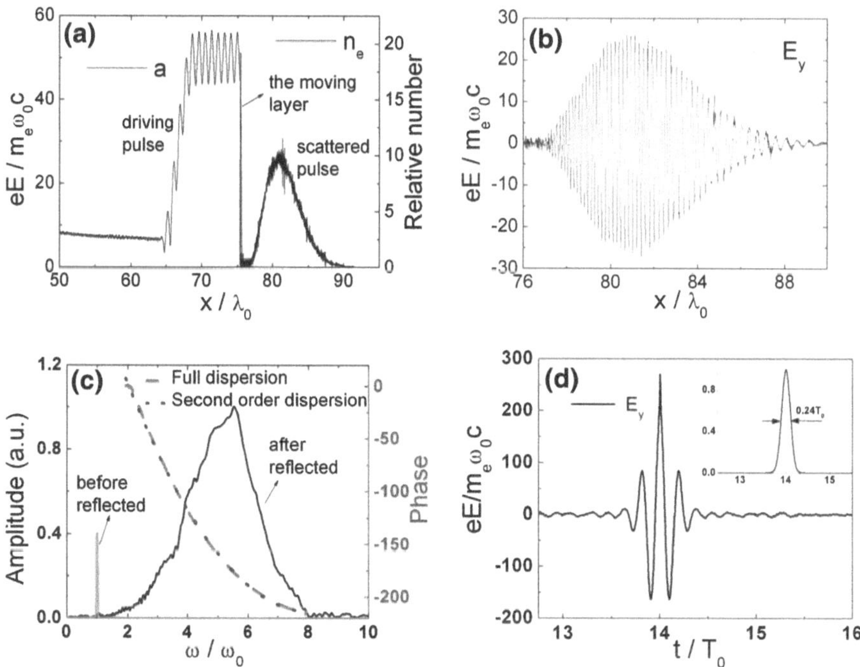

Fig. 5.1 Simulation results when the interaction is completed with the driving pulse of $a_0 = 50$, the scattered pulse of $a_{sc} = 5$, $\tau_{sc} = 30T_0$ and the foil of $n_0 = 100n_c$ and $d = 0.2\lambda_0$, respectively. **a** Laser field and electron density n_e distributions. **b** Transverse laser field E_y of the reflected chirped pulse. **c** Spectrum and phase of the scattered pulse: phase including full dispersion (*magenta-dashed*) and phase with only second-order dispersion (*blue-dotted*). The cyan solid shows spectrum of the scattered pulse before reflection. **d** Transverse laser field E_y and laser intensity profile after compression

5.3 Analysis

5.3.1 Characteristics and Manipulation of the Chirped Pulse

In this section, the mechanism will be analyzed in detail. Considering a laser pulse reflecting from a flying mirror, the Doppler Effect gives

$$\frac{\omega(t)}{\omega_0} = \frac{1 + \beta(t)}{1 - \beta(t)} = f(t) \tag{5.1}$$

where $\omega(t)$ is the temporal frequency of the reflected laser field, β is the layer velocity normalized by light speed c. Here f is defined as the ratio between the frequencies of reflected laser field and incident pulse. As the layer velocity increases, the reflected pulse gets chirped with time accompanying the frequency increasing. That is to say, the character of the reflected pulse is totally determined

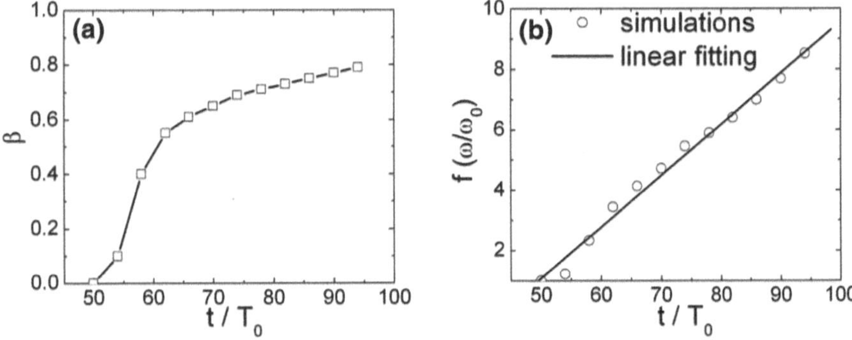

Fig. 5.2 **a** Velocity of the flying layer versus time. **b** The induced frequency shift increment factor f versus time. The velocity is obtained from simulation and the factor f is calculated according to Eq. (5.1)

by the layer motion. In Fig. 5.2 the layer velocity versus time and the corresponding frequency increment calculated from Eq. (5.1) are displayed. It is clearly seen that the layer velocity evolves in such a way that an optimistic linear frequency increment is induced on the reflected pulse, which is consistent with Fig. 5.1c. Since most pulse compression techniques require linearly chirped pulse, this favorable character should be very useful and convenient for future experiments.

In the above simulation, the SC pulse reaches the flying layer at about $t = 50T_0$ and leaves at $t = 90T_0$. During this period, the whole foil is accelerated by the DR pulse to a velocity of 0.77, as seen in Fig. 5.2a, which offers a frequency region of about $\omega_0 \sim 8\omega_0$ according to Fig. 5.2b. The peak laser field interacts with the layer at about $t = 70T_0$, corresponding to a central frequency of $5\omega_0$. These results are in perfect coincidence with Fig. 5.1c. The results in Fig. 5.2b provide a way of controlling the spectrum of the reflected pulse—choosing the appropriate interacting window. One could change the impinging time and/or pulse duration to control the low and/or high frequency bounds.

In addition, the features of the generated spectrum are controllable. According to Eq. (5.1), the instantaneous frequency of the chirped pulse can be well defined by varying the layer velocity. In LPA scheme, the layer motion is described as [2]

$$\frac{d}{dt}(\gamma\beta) = \frac{1}{2\pi m_i n_0 dc}\frac{1 - \beta}{1 + \beta}E_L^2(t - x/c). \tag{5.2}$$

Here m_i is the ion mass and $\gamma = (1 - \beta^2)^{-1/2}$ is the relativistic factor of the layer. As electrons and ions are accelerated as a whole, the longitudinal momentum of electrons is ignored. So, β can be expressed as a function of ω from Eq. (5.1) and we insert it into Eq. (5.2), thus gaining the form of the required laser electric field

$$E_L^2(t - x/c) = 2\pi m_i n_0 dc\frac{f + 1}{4f^{1/2}}\frac{df}{dt}, \tag{5.3}$$

$$\frac{x}{c} = \int_0^t \beta(t^{'})dt^{'}. \tag{5.4}$$

To produce a laser pulse with certain chirp, one may choose a proper profile for the DR laser following Eqs. (5.3) and (5.4). For example, if a linearly chirped pulse is acquired, f should take the form of $1 + kt$, and the calculated laser field is

$$E_L(t) = \sqrt{2\pi m_i n_0 dc}\sqrt{\frac{k}{2}}\left(\frac{e^{kt}}{2e^{kt/2} - 1}\right)^{1/4}. \tag{5.5}$$

Equation (5.5) shows accurate laser field form for generating an exactly linearly chirped pulse, while in the above simulation the laser field has a plateau profile such that chirp is only roughly linear in Fig. 5.2b.

5.3.2 Parametric Conditions

The parameters we used are chosen according to several requirements. First, the layer motion should not be affected by the SC pulse during interaction. This requires the light pressure of SC pulse $P_{sc} \sim a_{sc}^2(1+\beta)/(1-\beta)$ shall be ignored compared to that of the DR pulse $P_{dr} \sim a_0^2(1-\beta)/(1+\beta)$. Hence it gives the following relationship:

$$\frac{a_{sc}}{a_0} < \frac{1-\beta}{1+\beta}. \tag{5.6}$$

In the above simulation, $\beta_{max} \approx 0.77$ and we have $a_{sc} < 6.5$, which is well fulfilled.

Another important condition concern is with the two pulses interacting with each other in the confined electron layer. It was first studied by Shen et al. in Ref. [3] that if the two CP pulses with the same polarization direction overlap each other, high harmonics will be generated. To avoid the unfavorable effect, we demonstrated that the foil thickness should be larger than a critical value D (will be presented elsewhere)

$$d > D \approx \frac{a_0 + a_{sc}}{n_0}\frac{\lambda_0}{\pi} \tag{5.7}$$

D is calculated to be about 0.18 μm under the employed laser plasma parameters, thus we chose $d = 0.2$ μm. One can also avoid this effect by altering the polarization direction [3]. Besides, considering Doppler shifting the foil density should be high enough to stay opaque for the SC pulse, hence it gives $n_0 > f_{max} n_c$. In the above simulation $f_{max} \sim 10$, and the chosen parameter $n_0 = 100 n_c$ fulfills the requirement adequately.

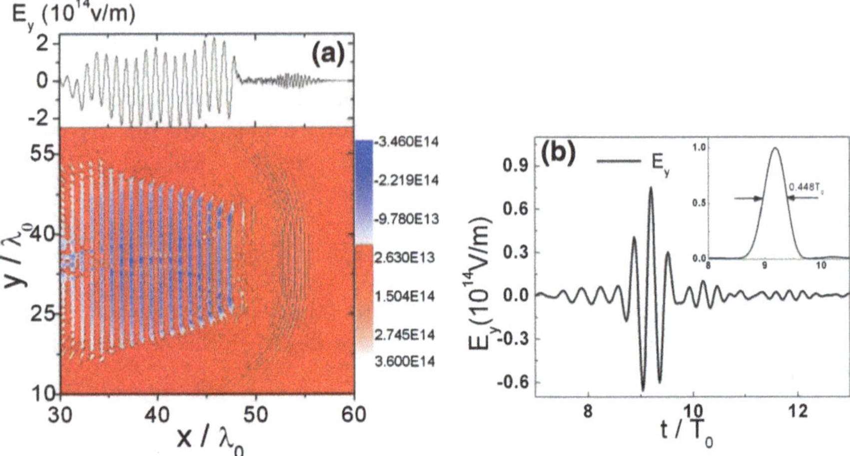

Fig. 5.3 **a** Distribution of transversal laser field E_y. **b** Transversal laser field E_y and laser intensity after compression

5.4 Two-Dimensional Simulation

As a routine, a multidimensional simulation is always necessary to show the validness of the proposal. In this case, a two-dimensional (2D) simulation is run in a box of $70\lambda_0 \times 70\lambda_0$, with cells of 3500×700. The amplitude of the DR pulse rises to $a_0 = 60$ in $2T_0$ and remains constant temporally. The SC pulse is with peak amplitude $a_{sc} = 5$ and duration $\tau_{sc} = 10T_0$. Both pulses are transversely super-Gaussian, $\boldsymbol{a} = a(t)\exp(-y^4/w_y^4)[\sin(\omega_0 t)\hat{e}_y + \cos(\omega_0 t)\hat{e}_z]$, where $w_y = 15\lambda_0$ for DR pulse and $16\lambda_0$ for SC pulse. The foil density is $n_0 = 40n_c$ and hence we choose its thickness as $d = 0.6\lambda_0$ according to Eq. (5.6). The foil is located at $34\lambda_0 - 34.6\lambda_0$.

Due to multidimensional effects, such as hole-boring and Rayleigh–Taylor (RT) like or Weibel-like instabilities [4–7], the foil accelerated in RPA scheme might be deformed therefore harmful for reflecting. To depress these effects, super-Gaussian laser pulse is employed in the simulation. One could also taper the foil to compensate the deformation [8]. Compared with 1D simulation, some of the parameters are also adjusted to give an optimized performance. The peak amplitude of the DR pulse is increased to $a_0 = 60$ and foil density is reduced to $n_0 = 40n_c$.

As is known, the instabilities develop much slower when the interface moves faster. Since the laser-driven hole-boring velocity scales as $a_0/n_0^{1/2}$ [5], the present parameters would offer a much larger velocity than that of 1D simulation and thus may better restrain the instabilities. Figure 5.3a shows the laser field distribution after interaction. A chirped pulse is well generated as a result of our modification. After compensating second-order dispersion, a near single-cycle laser pulse is obtained despite all the unfavorable effects. Its duration is about $0.448T_0$ and the wavelength is shortened to 0.32 μm, as displayed in Fig. 5.2b. To conclude, this mechanism works very well in multidimensional geometry.

5.5 Discussion

The proposed scheme is essentially a plasma approach, thus presenting no limit for laser intensity. An alternative way to compress pulse through laser-plasma interaction is the pure Doppler shifting method. The laser wavelength is decreased through Doppler shift after being reflected from a flying layer, while cycle number of the pulse remains the same. This method seems much simpler because no chirped pulse compression techniques are required. Nonetheless, our proposal shows several advantages. First, for the pure Doppler shifting method the dimensionless laser amplitude does not change after reflection, while in our proposal it is greatly enhanced after being compressed by optical approaches. Second, the spectrum of generated pulse with the pure Doppler shifting method is monochromatic. In our case, we consider furthermore the frequency increment induced by the acceleration of the layer. Therefore, the spectrum of the chirped pulse is highly broadened and no longer monochromatic. Lastly, the required layer velocity to obtain a commensurate short pulse is much smaller. As a comparison, to compress a 30 fs (normally used) laser pulse to about 1 fs, pure Doppler shifting method requires a moving layer with proton energy up to 3 GeV, while using this relativistic chirped pulse compression, as $\beta_{max} \approx 0.77$, one only needs proton energy of about 500 MeV.

5.6 Summary

In conclusion, combining Doppler shifting and dispersion compensation technique, we have proposed a method to generate relativistic single-cycled pulse with short wavelength by laser–foil interaction. A driving pulse accelerates a foil to form a fast moving layer and a counter-propagating SC pulse impinges on it. After being reflected from the layer, the SC pulse is Doppler shifted and strongly chirped. Applying dispersion compensation on the chirped pulse it can be compressed to nearly single-cycled relativistic pulse. This method has no intensity limit it only involves laser-plasma interaction. The dimensionless amplitude of the generated pulse is greatly increased and the wavelength is much shortened. Furthermore, the frequency characteristic can be determined by designing appropriate DR laser form, which is very convenient for application. We also examined the conditions that should be fulfilled to make this method efficient. Two-dimensional simulations are performed to check the validity and it shows that the proposal also works very well.

References

1. C. Nieter, J.R. Cary, J. Comp. Phys. **196**, 448 (2004)
2. T. Esirkepov et al., Phys. Rev. Lett. **92**, 175003 (2004)
3. B. Shen et al., Phys. Plasmas **8**, 1003 (2001)
4. B. Shen et al., Phys. Rev. E **64**, 056406 (2001)
5. S.C. Wilks et al., Phys. Rev. Lett. **69**, 1383 (1992)
6. A. Macchi et al., Phys. Rev. Lett. **94**, 165003 (2005)
7. M. Chen et al., Phys. Plasmas **15**, 113103 (2008)
8. X.Q. Yan et al., Phys. Rev. Lett. **103**, 135001 (2009)
9. M. Chen et al., Phys. Rev. Lett. **103**, 024801 (2009)

Chapter 6
Extreme Light Field Generation III: Ultra-Intense Isolated Attosecond Pulse

6.1 Introduction

In Chap. 1, the important applications together with the generating methods of attosecond (AS) pulses have been introduced. Attosecond pulses (APs) are usually obtained by lasers of moderate intensity interacting with atoms. However, the obtained light pulses are of very low intensity because the power of the driving ultra-short few-cycle laser pulse has to be kept low and the conversion efficiency ($10^{-7} \sim 10^{-6}$) is extremely small. In addition, a single-cycle driving pulse with appropriate carrier-envelope phase (CEP) [1, 2] or polarization gating technique [3] are required to obtain an isolated single AS pulse.

Another approach uses a relativistic linearly polarized (LP) laser pulse impinge on a solid target. The plasma boundary oscillates under the laser poderomotive force and then reflects the incident pulse. As the oscillating velocity approaches light speed, a bunch of high-order harmonics will be produced due to the relativistic Doppler effect. In the time domain, the reflected pulse exhibits a chain of intense APs. This is known as the "relativistic oscillating mirror (ROM)" model [4–11]. In ROM scheme the pulse intensity and efficiency are both greatly increased. Unfortunately, the nature of the ponderomotive force of LP pulse, which will be discussed later, suggests that the obtained APs are in a train while singe AP is a must for most proposed applications.

This chapter shows by particle-in-cell (PIC) simulations and analytical modeling that a few-cycle relativistic CP laser pulse impinging on an overdense plasma can produce a single AP by itself, with peak intensity higher than 10^{21} W/cm^2. The proposed scheme relies on the Doppler effect produced high harmonics in the reflected light when the CP laser pulse interacts with the laser-induced one-time oscillating boundary of the overdense plasma. The decay power law of its spectrum is consistent with the one predicted by the "ROM" model. The scheme has no limit on intensity caused by atom ionization threshold and does not require a stringent one-cycle diver pulse, nor polarization gating technique and precise follow-up frequency treatment.

L. Ji, *Ion Acceleration and Extreme Light Field Generation Based on Ultra-short and Ultra-intense Lasers*, Springer Theses,
DOI: 10.1007/978-3-642-54007-3_6,

6.2 CP Laser Field Reflected by Plasma Boundary

The features of reflected laser field from an oscillating plasma boundary is first presented via the 1D PIC simulation code LPIC ++. The simulation box is $10\lambda_0$ long with a resolution of $\lambda_0/1000$, where $\lambda_0 = 1\ \mu\mathrm{m}$ is the laser wavelength. A relativistic CP pulse with the form of $\boldsymbol{a} = e\boldsymbol{E}_L/m_e\omega_0 c = a_0 \sin^2(\pi t/2\tau_0)$ $[\sin(\omega_0 t)\hat{y} + \cos(\omega_0 t)\hat{z}]$, where e and m_e, $\boldsymbol{E}_L$ and ω_0 are electron charge and mass, laser electric field and frequency, respectively, c is the light speed, $a_0 = 20$ (or $I_0 = 1.1 \times 10^{21}\ \mathrm{W/cm^2}$), and $\tau_0 = 2\mathrm{T}_0$ (or 6.7 fs), is incident on an overdense plane target. The latter is fully pre-ionized and the ion charge and mass numbers are Z = 1 and A = 2, respectively. It is located in $6 < x\,[\mu\mathrm{m}] < 8$ and the density is $n_0 = 8n_c$, where $n_c = m_e\omega_0^2/4\pi e^2$ is the critical density.

The CP pulse impinges on the overdense target and is reflected. From the recorded electric field E_y in Fig. 6.1a one sees that the reflected pulse is clearly chirped, with its trailing part significantly steepened. Its spectrum shown in Fig. 6.1b, unlike the discrete or modulated one from LP laser interaction with the atoms in gases or solids, is continuous and smooth. This is due to a unique feature of CP laser interaction with overdense plasma, as shall be discussed later. It is crucial to the present scheme and is also the reason why the existing schemes for AP generation require elaborate frequency selection. After simple $\omega \leq 3\omega_0$ filtering of the low-frequency components, a single ultra-intense AP as short as 60 is obtained, as seen in Fig. 6.1a. The analytical model (see in Sect. 6.2) also predicts quite the same phenomenon in Fig. 6.1b. That is, a relativistic short CP pulse interacting with overdense plasma can generate a single AS light pulse naturally.

6.3 Analysis

6.3.1 Interacting Model

The analytical model of a CP laser pulse interacting with an overdense plasma is shown in Fig. 6.2 [12]. Plasma electrons are snow-ploughed by the laser ponderomotive force and pile up to form a dense skin layer |AB| with density of n_{p0}. The position A(x) (the origin is at O) denotes the displacement of the surface (or position of the “reflecting mirror”) of the electron layer. The momentum of an electron at the plasma surface is determined by the imbalance between the ponderomotive and space-charge forces

$$\frac{dp_x}{dt} = F_{px} - eE_{x0}, \tag{6.1}$$

where $E_{x0} = 4\pi n_0 e x$ is the charge separation field and F_{px} is the ponderomotive force experienced by the electron at x.

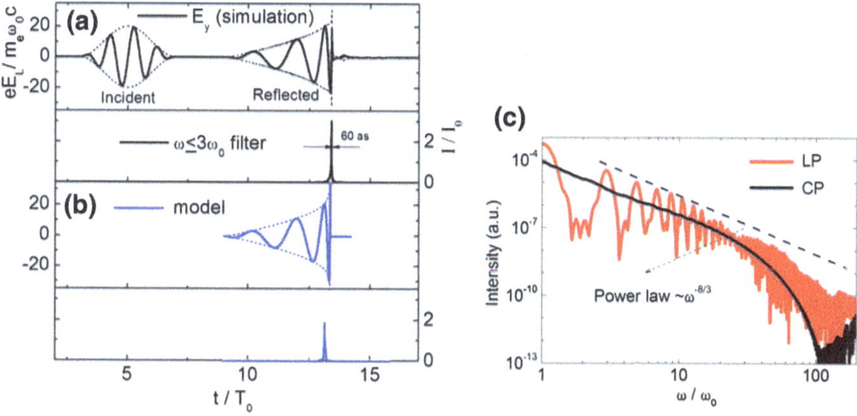

Fig. 6.1 Simulation and analytical results for the reflected light for an incident CP pulse with $a_0 = 20$, $\tau_0 = 2T_0$, and $n_0 = 8n_c$. **a** The transverse electric field E_y (*black solid*) and its envelope (*black-dashed*) at $x = 3$ μm of the reflected light. The panel below it shows the temporal intensity of the as pulse after the $\omega \leq 3\omega_0$ frequency filtering. Here $I_0 = 1.1 \times 10^{21}$ W/cm^2. The laser intensity is calculated by $I \sim E_y^2 + E_z^2$. **b** The corresponding results (*blue solid*) from the analytical model. **c** The spectra of the reflected light produced by CP and LP pulses. The *black-dashed* line is the power law predicted by the "ROM" model

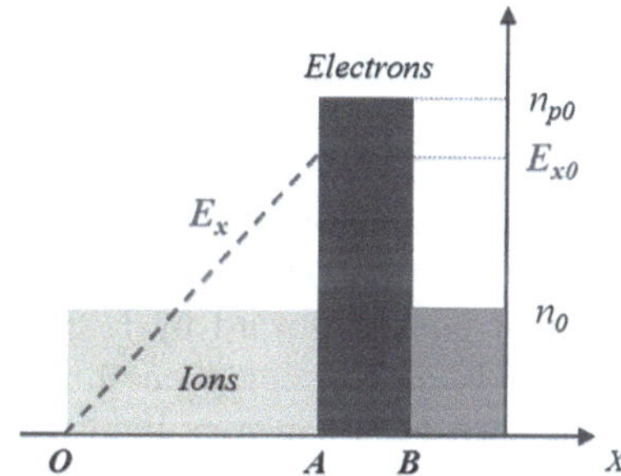

Fig. 6.2 Model of a CP laser pulse interacting with an overdense plasma

The ponderomotive force of the driver laser acts on the compressed electron layer AB of thickness equaling the skin depth $\lambda_s = c\sqrt{m_e/4\pi n_{p0}e^2}$. Assuming that it peaks at the boundary A and decays linearly to zero at B, we can write F_{px} as

$$F_{px} = \frac{4I(t - x/c)}{cn_{p0}\lambda_s}\frac{1 - \beta_x}{1 + \beta_x}, \tag{6.2}$$

where $\beta_x = dx/cdt$ and the Doppler effect has been taken into account. The intensity of the CP pulse is $I = \pi m_e^2c^5a^2/e^2\lambda_0^2$, where $a = a_0\sin^2[\pi(t - x/c)/2\tau_0]$. From charge conservation one obtains $n_{p0}\lambda_s = n_0(x + \lambda_s)$. Thus, Eq. (6.1) can be written as

$$\frac{d}{d\tau}(\gamma_x\beta_x) = \frac{8a^2}{N_0\left(\sqrt{X^2 + 1/\pi^2 N_0} + X\right)}\frac{1-\beta_x}{1+\beta_x} - 4\pi^2 N_0 X, \tag{6.3}$$

where $\gamma_x = (1-\beta_x^2)^{-1/2}$, $X = x/\lambda_0$, $\tau = t/\mathrm{T}_0$, and $N_0 = n_0/n_c$. The corresponding initial conditions are $X(0) = 0$ and $dX/d\tau|_{\tau=0} = 0$. At plasma boundary the reflected laser field and incident laser field give [9]

$$E_r(\tau + X) = -\frac{1-\beta_x}{1+\beta_x}E_i(\tau - X). \tag{6.4}$$

where $E_i(\tau - X) = E_0 \sin^2[\pi(\tau - X)/2\tau_0]$. Boundary motion can be numerically calculated from Eq. (6.3) together with initial conditions. The reflected laser field is obtained by Eq. (6.4).

6.3.2 Comparison with Simulations

The displacement of the target surface and velocity can be numerically obtained by substituting the parameters in Sect. 6.2 into Eq. (6.3), which is compared in Fig. 6.3 with the electron trails from 1D PIC simulation. They show excellent agreement between each other.

In the "ROM" model one usually does not expect the generation of high harmonics and APs by normally incident CP lasers, since with normally used parameters electrons in the plasma boundary are always steadily pushed forward by CP laser rather than oscillate as in the LP laser case. However, for the parameters employed above, from the electron trajectories in the simulation and the analytical model in Fig. 6.3, it is shown that the electrons at the target surface are first pushed inward by laser ponderomotive force and then they intensively bounce back, forming a one-time oscillation. The oscillation is the result of the imbalance between laser light pressure and space-charge separation field force. It is so intense that the maximum speed of the plasma boundary $-\beta_x$ (=0.996 at about $t = 10T_0$) can reach nearly the light speed. Such a high-speed reflecting mirror can severely compress the laser pulse and generate very high-order harmonics in the reflected light pulse, which results in an ultra-intense single AP in the time domain.

6.3.3 Effects of Target Density and Pulse Duration

The truth that the parameters employed above could induce such a drastic one-time oscillation and hence a single AP lies in two important facts. First the target density is close to the relativistic transparency threshold, in this case $n_{thr} \sim 5n_c$ [13].

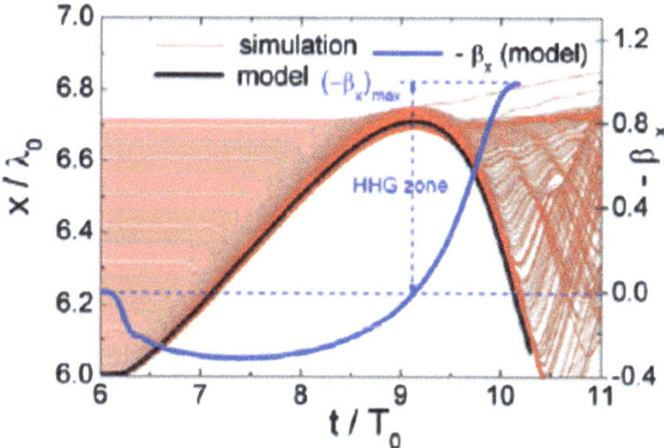

Fig. 6.3 Simulation and analytical results of the reflected light for an incident CP pulse. Electron trajectories (*thin red*) from the PIC simulation for the same parameters (see text) as in Fig. 6.1, and the displacement (*black solid*) and speed $-\beta_x$ (*blue solid*) of the electron surface layer as calculated from the analytical model. High-order harmonics are generated when $-\beta_x > 0$ (or $\omega_r > \omega_0$), marked as "HHG zone"

Instead of transmitting the target, the relativistic CP pulse snow-ploughs electrons to form a much denser layer in front of the interface, which would prevent further penetration and reflect laser field completely. The closer the target density to the relativistic transparency threshold, the higher the boundary oscillating velocity can reach and therefore the more intense the generated AP. Simulations and analysis indicate that as the density increases, peak intensity of the AP and the calculated peak value of the factor of the Doppler effect $\alpha = (1 - \beta_x)/(1 + \beta_x)$ both drop rapidly. This trend can be clearly seen in Fig. 6.4a.

Second, the pulse duration of the relativistic CP laser is much smaller than the ion responding time ($\sim 2\pi\omega_{pi}^{-1} \sim 20\mathrm{T}_0$). This guarantees that ions almost stay still during the whole interaction. Otherwise electrons will move together with ion flux and will not bounce back in an ultra-short time. If the duration is comparable to the plasma oscillating period, say $\mathrm{T_p/T_0} = (n_0/\gamma_0 n_c)^{-1/2} \sim 1.5$, here $\gamma_0 \sim (1 + a_0^2)^{1/2}$, an enhancement transportation of energy between incident laser and plasma boundary is stimulated. This would generate the most intense oscillation and hence the strongest AP. The peak intensity and $\alpha_{\max}$ exhibit similar trends that they approach maximum when pulse duration is comparable to plasma oscillating period and then drop with increasing duration, as seen in Fig. 6.4b. A single-period relativistic CP laser can generate a 40 as light pulse with a peak intensity $6.3 \times 10^{21}\mathrm{W/cm^2}$, which is a factor 6 more intense than the incident pulse.

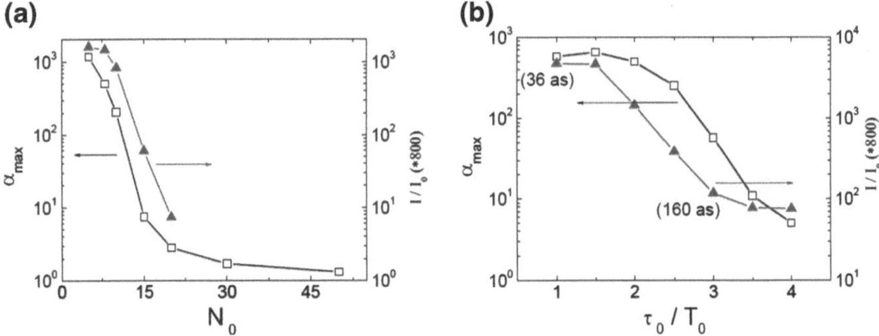

Fig. 6.4 Effects of the plasma density and pulse duration of the incident laser. **a** The normalized peak frequency α_{max} (*black squared*) from the model and the peak intensity (*red triangled*) from the simulations of the reflected light versus the target density. The above are for $a_0 = 20$ and $\tau_0 = 2T_0$. **b** The normalized peak frequency from the model and the peak intensity from the simulations versus the pulse duration, for $a_0 = 20$ and $n_0 = 8n_c$

6.3.4 Comparison with LP Laser

In the ROM scheme, normally LP lasers are used to produce APs. It is instructive to compare it with our proposal. For the simulation, the LP pulse takes the form of $\hat{a} = a_0 \sin^2(\pi t/2\tau_0)\sin(\omega_0 t)\hat{y}$, where $a_0 = 20\sqrt{2}$ and $\tau_0 = 2T_0$. The plasma density is $n_0 = 100n_c$ (commonly used for LP pulses). From the recorded laser field in Fig. 6.5a, one sees that the laser pulse is also reflected by the oscillating electron layer and significantly reshaped, with several strongly steepened edges. After the $\omega \leq 3\omega_0$ frequency filtering each steepened edge in the high-intensity region results in an AP. Four intense APs are generated.

The differences in the behavior of the reflected lights from the LP and CP pulses are mainly due to a difference in the property of the overdense plasma boundary during the interaction. When the target is illuminated by a laser pulse, the plasma surface experiences a ponderomotive force

$$F_P = \begin{cases} -\dfrac{e^2}{8m_e\omega_0^2}\nabla E_0^2(x)(1-\cos(2\omega_0 t)), & \text{for LP pulse;} \\ -\dfrac{e^2}{4m_e\omega_0^2}\nabla E_0^2(x), & \text{for CP pulse.} \end{cases} \tag{6.5}$$

where $E_0(x)$ is the envelope of the laser field. For the LP pulse, the ponderomotive force contains a $2\omega_0$ component, so that the plasma boundary oscillates twice per laser period, or four per pulse for a two-cycle laser. The consequence is clearly exhibited in the electron trajectories that there are four, instead of one as in the CP driver case, turns in each trajectory. According to the "ROM" mechanism, each turn of the electrons in the space-time domain generates a spike in γ [9] and thus high harmonics in the reflected light. Accordingly, the latter still contains four APs after the $\omega \leq 3\omega_0$ frequency filtering. Unlike the CP driver case, it is not possible

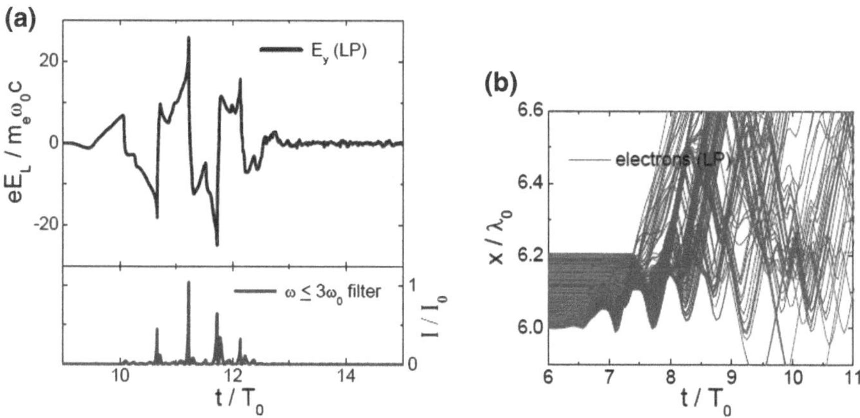

Fig. 6.5 Simulation results for an LP driver pulse with $a_0 = 20\sqrt{2}$, $\tau_0 = 2T_0$, and plasma density $n_0 = 100n_c$. **a** The light electric field E_y at $x = 3\,\mu$m (*black solid curve* in the *upper panel*) and the light intensity after $\omega \leq 3\omega_0$ frequency filtering (*blue solid curve* in the *lower panel*). **b** Electron trajectories from the simulation

to produce a single AS pulse when a normally incident LP driver laser is used. Since even a single-cycle LP driver pulse would still generate two APs. The spectra of the reflected lights of the CP and LP drivers are both consistent with the power law $I_\omega \sim \omega^{-8/3}$ predicted by the "ROM" theory. As discussed, a key difference between the two is that the spectrum from the CP driver is continuous, and that from the LP driver is highly modulated and discrete. The modulation is due to the abrupt changes in the electron γ arising from the increased number of plasma boundary oscillations within the driver-pulse period [2].

Since it comes to few-cycle laser pulse regime, the effects of phase offset between laser field and carrier envelope should be addressed. The generated APs of CP and LP lasers with three phase offsets $\pi/4$, $\pi/2$ and π are shown in Fig. 6.6d–i, respectively. One can see that in our proposal, i.e., using CP laser pulse, the generated single ultra-intense AP is irrelevant to phase offset of the incident laser, where three identical single APs are produced under different phase offsets. It is very favorable for many applications. On the contrary, the results of using LP laser strongly rely on initial phase offset. They all generate two APs. Nevertheless, Fig. 6.6h also indicates that using one-cycle LP laser with certain phase offset an isolated intense AP might be obtained after appropriately filtering the much weaker one.

6.4 Two-Dimensional Simulation

In this section, a two-dimensional (2D) PIC simulation of the proposed scheme was carried out by the code VORPAL. The size of the simulation box is $15\lambda_0 \times 30\lambda_0$ and there are 4500×1200 cells in the x and y directions, respectively.

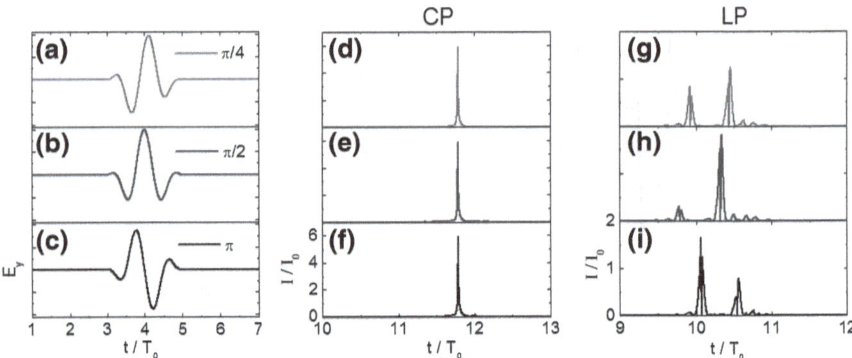

Fig. 6.6 Effects of phase offset in the incident pulse on generated APs. The parameters are $a_0 = 20$, $\tau_0 = \mathrm{T}_0$ and $n_0 = 8n_c$ for CP laser and $a_0 = 20\sqrt{2}$, $\tau_0 = \mathrm{T}_0$ and $n_0 = 100n_c$ for LP laser, respectively. The incident light field E_y with phase offset $\pi/4$ (**a**), $\pi/2$ (**b**) and π (**c**); generated APs under three different phase offsets with CP laser (**d**)–(**f**) and LP laser (**g**)–(**i**). All APs are obtained after $\omega \leq 3\omega_0$ frequency filtering

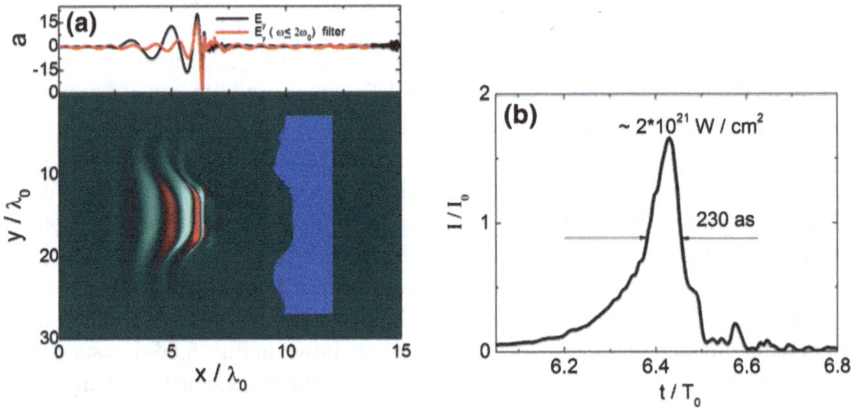

Fig. 6.7 Results from the 2D PIC simulation with the same laser and plasma parameters as that in Fig. 6.1. **a** Reflected light field E_y. The *upper panel* shows the axial profile of E_y at $y = 15\,\mu\text{m}$ and the temporal electron density distribution. **b** Intensity profile of the generated AP

A target of density $n_0 = 8n_c$ is located in the region $10 - 12\,\lambda_0$ in the x direction. The CP pulse is the same as that used in Fig. 6.1 except that it is transversally super-Gaussian. The transverse electric field E_y distribution displayed in Fig. 6.7a shows that the laser pulse is well reflected from the overdense plasma layer. As in the 1D case, the trailing edge of the reflected pulse is highly compressed to form a peak, which contains high-order harmonics. After filtering the fundamental and diploid frequency components, a single ultra-short light pulse is obtained. The AP has a duration of 230 as and its peak intensity reaches $2 \times 10^{21}\,\text{W/cm}^2$.

The conversion efficiency is somewhat more than 1 %, or 4 order of magnitudes higher than that from laser–atom interactions.

As is well known, several multidimensional instabilities can occur in laser–plasma interaction [14–16], and these instabilities may deform or destroy the reflecting plasma surface. However, in the 2D simulation no evidence of any instability was found. This is due to the fact that the timescale involved is so short that instabilities have no time to develop. In fact, with the single-cycle driver [3] the generated single AP is almost the same in the 1D and 2D geometries.

6.5 Summary

In conclusion, a plasma approach for generating isolate ultra-intense AP is proposed. When a relativistic CP pulse impinges on an overdense plasma, the plasma boundary would oscillate and reflect the incident light. Due to the Doppler effect, the reflected pulse contains high-order harmonics and its spectrum is continuous. An isolated AP of light can be obtained after simple frequency filtering. The results agree well with an analytical model for the interaction process. Simulations and model explained why such remarkable AP can be generated using CP laser. They also show the effects of pulse duration and target density. Unlike the traditional method of AP generation by laser–atom interaction, the proposed scheme has no intensity limit for the driver laser and the conversion efficiency is much higher. In addition, it does not require a strict one-cycle laser.

References

1. G. Sansone, E. Benedetti, F. Calegari, C. Vozzi, L. Avaldi, R. Flammini, L. Poletto, P. Villoresi, C. Altucci, R. Velotta, S. Stagira, S. De Silvestri, M. Nisoli, Science **314**, 443 (2006)
2. E. Goulielmakis, M. Schultze, M. Hofstetter, V.S. Yakovlev, J. Gagnon, M. Uiberacker, A.L. Aquila, E.M. Gullikson, D.T. Attwood, R. Kienberger, F. Krausz, U. Kleineberg, Science **320**, 1614 (2008)
3. X. Feng, S. Gilbertson, H. Mashiko, H. Wang, S.D. Khan, M. Chini, Y. Wu, K. Zhao, Z. Chang, Phys. Rev. Lett. **103**, 183901 (2009)
4. S.V. Bulanov, N.M. Naumova, F. Pegoraro, Phys. Plasmas **1**, 745 (1994)
5. R. Lichters, J. Meyer-ter-Vehn, A. Pukhov, Phys. Plasmas **3**, 3425 (1996)
6. S. Gordienko, A. Pukhov, O. Shorokhov, T. Baeva, Phys. Rev. Lett. **93**, 115002 (2004)
7. U. Teubner, K. Eidmann, U. Wagner, U. Andiel, F. Pisani, G.D. Tsakiris, K. Witte, J. Meyer-ter-Vehn, T. Schlegel, E. Förster, Phys. Rev. Lett. **92**, 185001 (2004)
8. U. Teubner, P. Gibbon, Rev. Mod. Phys. **81**, 445 (2009)
9. T. Baeva, S. Gordienko, A. Pukhov, Phys. Rev. E **74**, 046404 (2006)
10. B. Dromey, S. Kar, C. Bellei, D.C. Carroll, R.J. Clarke, J.S. Green, S. Kneip, K. Markey, S.R. Nagel, P.T. Simpson, L. Willingale, P. Mckenna, D. Neely, Z. Najmudin, K. Krushelnick, P.A. Norreys, M. Zepf, Phys. Rev. Lett. **99**, 085001 (2007)

11. B. Dromey, M. Zepf, A. Gopal, K. Lancaster, M.S. Wei, K. Krushelnick, M. Tatarakis, N. Vakakis, S. Moustaizis, R. Kodama, M. Tampo, C. Stoeckl, R. Clarke, H. Habara, D. Neely, S. Karsch, P. Norreys, Nat. Phys. **5**, 146 (2009)
12. A. Macchi, F. Cattani, T.V. Liseykina, F. Cornolti, Phys. Rev. Lett. **94**, 165003 (2005)
13. B. Shen, Z. Xu, Phys. Rev. E **64**, 056406 (2001)
14. S.C. Wilks, W.L. Kruer, M. Tabak, A.B. Langdon, Phys. Rev. Lett. **69**, 1383 (1992)
15. M. Chen, A. Pukhov, Z.M. Sheng, X.Q. Yan, Phys. Plasmas **15**, 113103 (2008)
16. X.Q. Yan, H.C. Wu, Z.M. Sheng, J.E. Chen, J. Meyer-ter-Vehn, Phys. Rev. Lett. **103**, 135001 (2009)

Chapter 7
Summary

Researches on relativistic laser plasma interaction have made great progress in laser wakefield acceleration of electrons and laser-ion acceleration. Among the interactions, laser polarization plays an important role. Circularly polarized (CP) lasers, as its unique feature of the related ponderomotive force, show great advantages in light-pressure acceleration of ions, and thus are becoming more and more important. This book relies on the interaction of CP lasers and overdense plasmas, performing study of laser-ion acceleration and extreme light field generation by theoretical and simulative approaches.

The study of laser-ion acceleration contains two aspects:

1. Using particle-in-cell simulations to investigate the process of a laser interacting with a compound plasma target, which includes two ion species. The results revealed that both light and heavy ions are accelerated to the same velocity. The common velocity is higher than that with a pure heavy ion target. Hence acceleration of heavy ions can be enhanced by mixing them with light ions. A "sandwich target" structure is designed to further improve the energy spread of the beam. Three-dimensional particle-in-cell (PIC) simulation generated a 58 MeV carbon ion beam with only 5 % energy spread, by using a $5 \times 10^{19} \mathrm{W/cm^2}$ laser pulse. Related results were included in Chap. 2.
2. In Chap. 3, the critical thickness issue in light-pressure acceleration was addressed. Analysis and simulations indicate that the rising front of the laser pulse is crucial for the critical value. When the pulse rises gently, stable acceleration still survives and even the foil is much thinner than formerly predicted. Thinner foil means higher peak energy and efficiency.

Generation of extremely short and intense light fields includes three aspects:

1. A nonlinear modulation effect of foil transparency was discovered in relativistic laser–foil interaction in Chap. 4. When a CP laser impinges on a thin foil, the most intense part will penetrate, leaving the weaker front and tail reflected by

L. Ji, *Ion Acceleration and Extreme Light Field Generation Based on Ultra-short and Ultra-intense Lasers*, Springer Theses,
DOI: 10.1007/978-3-642-54007-3_7,

the foil. The duration of the transmitted pulse is greatly reduced. The mechanism can produce a 4 fs, $3 \times 10^{20}\,\mathrm{W/cm^2}$ near single-cycle laser pulse.

2. Dual-laser interacting with double-sided foil can generate a strong chirped pulse with broad-band spectrum. The foil driven by the driving laser serves as a flying mirror, by which the other pulse is scattered. Due to the acceleration of the foil, the scattered (SC) pulse is strong chirped. The broadband SC pulse forms a short wavelength single-cycle laser beam after its dispersion is fully compensated. Further focusing may boost the intensity to four magnitudes higher than the initial one. The mechanism was described in Chap. 5.
3. A method of generating single ultra-intense attosecond pulse (AP) based on laser–plasma interaction was presented in Chap. 6. The few-cycle relativistic CP laser can drive the plasma boundary to form a drastic one-time oscillation, where the incident pulse is reflected. Due to the Doppler effect, high harmonics and hence an isolated AP is induced to the reflected pulse. In simulations, a 36 as laser is generated with intensity of $6.3 \times 10^{21}\,\mathrm{W/cm^2}$.

Looking to the future, light-pressure acceleration of ions is the most promising scheme for producing GeV ions. Although a lot of progress has been made, a number of difficulties remain. Among them the most urgent ones are the instabilities. Application of laser-ion acceleration requires more stable schemes, which are definitely the upcoming concerns.

On the other hand, we have shown that in the relativistic region, plasma can serve as a powerful medium that sustains the generation of extremely short and intense light pulses. Several meaningful explorations are made in this book, and we hope that more attention will be paid to this exciting topic—plasma-based relativistic optics.

Zeitfracht Medien GmbH
Ferdinand-Jühlke-Straße 7
99095 Erfurt, Deutschland
produktsicherheit@kolibri360.de